Ultrasonic Measurement and Characterization of Liquid-Particle Flow

Barbara Omylska

Ultrasonic Measurement and Characterization of Liquid-Particle Flow

Petroleum Production Engineering with Data Science

Springer

Barbara Omylska
Tarnów, Lesser Poland, Poland

ISBN 978-3-032-03956-9 ISBN 978-3-032-03957-6 (eBook)
https://doi.org/10.1007/978-3-032-03957-6

This Springer imprint is published by the registered company Springer Nature Switzerland AG
The registered company address is: Gewerbestrasse 11, 6330 Cham, Switzerland

Foreword

This book is dedicated to the application of ultrasound for characterizing liquids when they flow, including complex liquids like particle suspensions. The idea of using ultrasound for this purpose is very old, dating back many decades. There have been hundreds, perhaps thousands of papers published on this subject during the last century. Many of them mentioned process control as the major justification of the study.

As a result of these efforts, we now have a widely accepted and well-verified theory. There have been three books published on this subject by Elsevier. The last one is: A.S. Dukhin and P.J. Goetz (2017). *Characterization of Liquids, Dispersions, Emulsions, and Porous Materials Using Ultrasound*, Elsevier, Third Edition.

The study by Barbara Omylska sheds lights on many aspects of pulsed broadband methods that were previously questionable and inconclusive, including the verification of their performance in terms of accuracy and precision in process applications, such as flowing liquid-particle suspensions. This broadband approach has attracted the attention of various groups from time to time, because it offers certain benefits for process control with potential advantages for optimizing many industrial applications.

This book is complementary to my own. Researchers will be able to make a proper selection for their potential projects.

Dispersion Technology, Inc.
Bedford Hills, NY, USA

Andrei Dukhin

Preface

A research project described in this book was initiated thanks to successful collaboration between the University of Stavanger in Norway and Total E&P Norge AS (the last-mentioned has become TotalEnergies EP Norge AS).

This innovation, research and development project consists of theoretical, experimental, data science, finite element modeling, mechanical design and project management components, for the purpose of description of liquid-particle flow using ultrasound techniques. The objective of the project was to develop the ultrasonic methodology for measurement and characterization of liquid flow containing solid particles, in online (real-time) measurement conditions.

The project consists of primary and complementary studies. The primary studies comprise development of the ultrasonic methodology for the simultaneous measurement of the average particle concentration and flow speed parameters. The performed studies also included the ultrasound sensitivity to flow turbulence intensity, the characterization of rheological properties of fluids, the detection of particle clustering and air bubbles as the third phase. The developed methodology is based on the ultrasonic attenuation spectroscopy and sound speed measurement methods, supported by data science programs. The objective of achieving online acoustic spectroscopy measurement of flowing particle suspensions is, in global terms, in an ongoing development phase.

The basic elements of the improved pulsed broadband ultrasonic attenuation spectroscopy method were studied: selecting and testing the experimental setup's elements in terms of their limitations and measurement sensitivity, designing an ultrasonic pulse that is optimal for the detection of a small particle and for the measurement of liquid-particle flow, signal acquisition techniques, signal processing, including averaging techniques and the time-to-frequency conversion, sound resonances.

Experimental results from the tests with water-glass particle suspensions of low particle concentrations in high-shear flow are presented. Since several effects were measured simultaneously by one ultrasonic pulse, the interpretation of the results is complex and requires programming codes as the support. The acoustic theories for particulates, the damped, driven harmonic oscillator, and the sediment transport theories, as well as published results and performed numerical simulations in MATLAB, bring together the necessary knowledge for the proper interpretation of the results.

Acoustic fields in liquid and liquid-particle suspensions were calculated in COMSOL Multiphysics (4D models), for better understanding of the experimental results and visualization of the ultrasound waves propagation.

The complementary studies include acoustic streaming, designing a small-scale rotational test cell with an annular flow channel, testing it for various applications and characterization of the flow regimes. Flow speed measurements were performed using three methods: the ultrasonic transit-time difference, the cross-correlation of the backscattered from particles ultrasonic echoes, the particle image velocimetry.

Applications of the project are related to monitoring of solids and liquid flow, control of flow dynamics parameters, such as particle concentration and flow speed, flow turbulence detection. These are petroleum production engineering applications, however, the developed methodology is suitable for a range of other industries.

Valuable insights came from the Institute for Multiphase Flow Research (IMFT) in Toulouse, France. Experiments were performed in the Multiphase Flow and Instrumentation Laboratory at the University of Stavanger, Norway. Related field experience was gained in Statoil's (Statoil has become Equinor) Research and Development Process Centre in Porsgrunn, Norway. The laboratory work experience related to this project was gained at the University of Science and Technology in Trondheim, Norway, while working on another project about ultrasonic measurement of falling liquid films at the University of Stavanger. Valuable knowledgeable support was received from Dispersion Technology, Inc. in New York, United States. Guidance and assistance with the book publishing were received from Springer and Springer Nature. The author expresses gratitude for all the received support.

Tarnów, Lesser Poland, Poland Barbara Omylska

Competing Interests The author has no competing interests to declare that are relevant to the content of this manuscript. Elsevier, Applied Acoustics and the Nordic Rheology Society are kindly acknowledged for giving permissions to publish the source material in this book.

Contents

About the Author

Barbara Omylska has a PhD in Petroleum Engineering from the University of Stavanger. She accomplished a Postgraduate Certificate in Project Management from the University of Liverpool. Her affiliations included majors in the energy and investment banking industries in Norway, the United Arab Emirates and Poland. She published her experience in engineering, science, and management, with the intention of supporting the world.

Chapter 1
Introduction

There is an underexplored area in the petroleum industry, related to online (real-time) measurement of liquid-solid particle flowing suspensions. The flow of a suspension is a challenge in the metering world, even though it commonly occurs in transport pipelines, multiphase separators, and other process facilities in the petroleum industry. Small solid particles tend to be easily suspended in the main liquid flow, which can cause damages to production facilities, when the process is uncontrolled. Therefore, there is a need to provide information for the control of flow dynamics parameters, in terms of particle concentration and flow speed. Other flow dynamics parameters, including flow turbulence intensity and particle size are also of interest, to fully describe two-phase liquid-particle flowing systems. Turbulent flow exists in 95% of closed pipelines, therefore, flowmeters should normally be designed to operate under such flow conditions. However, this is a challenge due to flow turbulence effects at high flow velocities.

Ultrasound technology is environmentally friendly, noninvasive, externally applied and well-proven in the medical service, the engineering industries and the physical sciences. Among several existing applications, some examples of medical applications are cancer treatments, breaking kidney stones, and detecting the heartbeat of a fetus. Industrial applications can be listed as thickness measurement, nondestructive testing (NDT), flow speed measurement, characterization of dispersions, manipulation of particles, enhancement of flow in microchannels, ultrasound imaging, real-time identification of moving objects. The ultrasound technology has been extensively developed to provide a detailed description of complex systems. The flow of liquid-solid particle suspensions is one of them.

This research project describes the development of the ultrasonic methodology for measurement and characterization of liquid flow containing solid particles (Wrobel 2012). Ideas were put into actions to answer the following questions: Which ultrasonic technique is suitable for the real-time measurement of liquid-solid particle flow and why? Which flow dynamics parameters can be described using ultrasound measurement methods? What is the effect of particles flowing in liquid

B. Omylska, *Ultrasonic Measurement and Characterization of Liquid-Particle Flow*, https://doi.org/10.1007/978-3-032-03957-6_1

on the ultrasonic measurement? Is it possible to measure flow speed and particle concentration simultaneously using the same type of ultrasonic pulse? What is the effect of flow turbulence on the ultrasonic measurement? These questions are the key questions shaping the content of this book.

1.1 Project Objectives

1.1.1 Technical Objectives

The main technical objective of this project is to develop the ultrasonic methodology for measurement and characterization of flow containing solid particles. A challenge is to measure particle concentration and flow speed simultaneously in online conditions.

The creation of the innovative methodology (the methods and processes containing technology) is based on the pulsed broadband ultrasonic attenuation spectroscopy (using multifrequency frequency time-domain signals) and the sound speed measurement methods.

Another objective is to explore possibilities of ultrasounds for the control of other flow dynamics parameters, for example, flow turbulence intensity and particle size.

The project's technical outcomes are the methodology, the advanced measurement system, the experimental results, the numerical examples, and the publications.

1.1.2 Project Management Objectives

Project management methods and tools enabled the project delivery. Agile, lean and waterfall (hybrid) management methods were utilized, as well as ongoing constructive reflection. The framework involved defining the project scope, requirements and limits, daily project planning (including a roadmap), the creation of the front-end strategy, the evaluation and selection of concepts, the development of the project tasks' activities and prioritization, the current status capture and decisions for the future, the integration of the subprojects into the complex long-term project, progress reporting, the monitoring of the project metrics, the conceptual-experimental-modeling knowledge implementation, continuous improvement, and the project delivery. Importance was also placed on the utilization of scientific standards, for example, ISO standards.

Relation management and cooperation with external institutions, writing, publishing and public speaking on the subject, as well as the experience of received mentorship and given coaching were inherent to the project.

1.2 Project Applications

The project's applications are related to the petroleum production engineering and flow assurance areas: hydrocarbons production process monitoring with respect to liquid with solids flow, sand production, the control of flow dynamics parameters, namely the concentration of solids and flow speed, the detection of instabilities in a turbulent flow regime, undesired flow regimes in a multiphase flow separator, other production and process facilities.

Ultrasonic flowmeters are the industrial applications of ultrasound technology. The project's results, knowledge and applications are valuable for a range of other industries.

1.3 A Project Structure

The project contains the following parts:

- An overview of available ultrasound methods for the characterization of liquid and liquid-particle flows.
- A discussion of the basic elements forming the foundation for development of the ultrasound methodology for measurement of liquid-particle flow.
- Introduction into the acoustic, physical, and fluid dynamic theories relevant for this project.
- A discussion of the primary study results. The primary studies comprise the ultrasonic methodology development for the ultrasonic measurement of liquid-particle flow, in terms of the simultaneous determination of average particle concentration and flow speed. It is based on the pulsed broadband ultrasonic attenuation spectroscopy and the sound speed measurement methods. The developed methodology is also useful for the liquid-particle flow characterization in terms of the detection of flow turbulence intensity, particle clustering, air bubbles as the third phase and for the study of the physical properties of fluids.
- A discussion of the results from the complementary studies:
  - The flow speed measurements of suspensions using the ultrasonic transit-time difference (TTD) method and verification using the particle image velocimetry (PIV) method.
  - The flow speed measurements of suspensions using the cross-correlation function (CCF) method of ultrasonic signals backscattered from solid particles.
  - Finite element modeling of acoustic fields in liquid and suspensions using COMSOL Multiphysics.
  - Designing a small-scale rotational flow cell, testing it for various flow regimes and characterization of particle suspensions.

- A list of main achievements.
- Conclusions and recommendations for future research.
- MATLAB programming codes for time series and images.

1.4 Innovation Project Metrics

Metrics are needed to avoid disappointments about uncompleted or completed projects. Their role is to actively protect, support and direct projects toward achieving success. They provide essential project information to bring the attention of stakeholders and for control. The most important metrics for research and development projects are adapted and extended, based on Kerzner's work (2013). The metrics of innovation projects enable flexibility, as well as discipline toward results delivery. The metrics are listed below:

- A complexity factor (related to the project length and uncertainty).
- Customer satisfaction rating (the input from customers shapes innovation projects).
- The number of critical assumptions (related to the project feasibility).
- The percentage of the critical assumptions that have changed (related to communication, stakeholder management and knowledge transfer in a project).
- The percentage of the work packages with critical designations (defines required focus on the specific project's components).
- The number of the scope change control meetings (related to the project feasibility under current limitations and the customers' expectations).
- The quality of work results.
- The percentage of the work packages on schedule (defines the project completion in different phases).
- The percentage of the work packages on budget (defines the range of incurred costs).
- Planning for the budget and expenditures incurred (defines the success of the project's completion within its budget).
- Milestones (they are needed for checking the health of innovation projects and it is best to define them in the initial project phase).
- Tangible early signs of success (related to the forecasting of the success rate).
- The amount of training provided (correlated with the project success).
- Team members' satisfaction rating (especially important for long-term innovation projects).
- The priority of activities, supported by the project risk matrix, for better allocation of resources.
- Effort (related to the work intensity input into a project).
- The number of excellence cases (points to the recognition of the project performance and the need for support and reinforcement).

Competing Interests The author has no conflicts of interest to declare that are relevant to the content of this chapter.

References

H. Kerzner, *Project Management Case Studies*, 4th edn. (Wiley, Hoboken, New York, 2013)
B.M. Wrobel, *Ultrasonic Measurement and Characterization of Liquid-Particle Flow*. PhD thesis no. 171, University of Stavanger, Norway, 2012

Chapter 2
The Ultrasonic Methodology Development for Liquid-Particle Flow

2.1 Ultrasonic Techniques for the Control of Flow Parameters

Several ultrasonic measurement techniques suitable for the control of flow parameters were evaluated, including their applications and measurement systems. The study themes in Tables 2.1, 2.2, 2.3, and 2.4 were chosen based on their relevance to the stated project scope in Sects. 1.1 and 1.2. Based on the findings, decisions were made on how to proceed with the experimental part of the project for building the ultrasonic methodology, and which project's tasks were given priority in the initial front-end stage.

Various ultrasonic techniques were studied regarding suitability and complexity (Tables 2.1, 2.2, 2.3, and 2.4), together with availability of resources for the project. The ultrasonic attenuation spectroscopy technique seemed to be the most optimal for the measurement of average particle concentration, using the through-transmission method (Theme 2 in Table 2.1). There is large potential for the detection of particle concentration and size using this technique. The goal is measuring different particle concentrations. The transmitted signal will be analyzed with respect to the time delay and amplitude damping. Ultrasonic direct-contact transducers having flat surfaces will be used. Different wave shapes (the sine, pulse, sweep, square, triangle) will be studied. An additional advantage of the chosen ultrasonic method is the possibility to study the phenomena numerically, for example, using the finite element-based acoustic module in COMSOL Multiphysics. The ultrasound energy can be calculated for various liquid-particle cases using the incident waves of different shapes and frequencies. The simulations can be performed for different particle concentrations and distributions.

The ultrasonic attenuation spectroscopy method could also be performed using the pulse-echo technique to analyze the frequencies of the reflected signal from a liquid-particle mixture (Theme 1 in Table 2.1). It is suspected that this method

B. Omylska, *Ultrasonic Measurement and Characterization of Liquid-Particle Flow*, https://doi.org/10.1007/978-3-032-03957-6_2

Table 2.1 Study themes for the ultrasonic measurements of liquid-particle systems.

No.	**1**
Study themes	Particle concentration and size parameters (Kobayashi et al. 2006).
Measured parameters	Amplitude (sound attenuation) and frequency of backscattered ultrasound waves.
An ultrasonic measurement system and a procedure	One transducer for the pulse-echo (PE) technique. Sending a sine wave to a system and measurement of the frequency response and reflected signal attenuation.
Ultrasonic and signal processing techniques	The ultrasound attenuation spectroscopy (time and frequency domains) using the PE technique and a fast Fourier transform for the signal conversion into the frequency.
Compatibility with other techniques	High-speed video and laser techniques. Finite element modeling using COMSOL Multiphysics.
No.	**2**
Study themes	Particle concentration, size, particle size distribution parameters (Beckmann et al. 2006). An idea for an initial experiment using the through-transmission (TT) technique.
Measured parameters	A time delay of a signal, amplitude attenuation. Measurement using the PE and TT techniques. The frequency spectrum of the reflected signal from particles. It is dependent on the particle concentration and frequency of the incident sound beam.
An ultrasonic measurement system and a procedure	One transducer for the PE mode. One emitting transducer and one or two receiving transducers for the TT mode. Possibly applying a moving transducer. Using sine waves of different pulse repetition frequencies (PRF) for testing. Additionally, using a broadband pulse. The possibility of designing a wave.
Ultrasonic and signal processing techniques	The time-domain spectroscopy (TDS).
Compatibility with other techniques	High-speed video and laser techniques. Finite element modeling.
No.	**3**
Study themes	The concentration and density of slurry systems.
Measured parameters	The sound speed of a mixture. Possibly also the sound attenuation.
An ultrasonic measurement system and a procedure	Applicable for homogenous mixtures.
Ultrasonic and signal processing techniques	Correlating the sound velocity and/or the signal attenuation. Measurement using the TT technique.
Compatibility with other techniques	
No.	**4**
Study themes	An analysis of an acoustic speckle signal.

Table 2.1 (continued)

Measured parameters	The spatial displacement of an object, the size of particles in fluid (Bristol General Hospital 1999).
An ultrasonic measurement system and a procedure	
Ultrasonic and signal processing techniques	The speckle pattern correlation method. The ultrasonic imaging methods.
Compatibility with other techniques	
No.	**5**
Study themes	A location determined by ultrasonic methods.
Measured parameters	The time delay between the transmitted and reflected signals.
An ultrasonic measurement system and a procedure	One ultrasonic transducer.
Ultrasonic and signal processing techniques	The PE technique.
Compatibility with other techniques	
No.	**6**
Study themes	Velocity profile (Ozaki et al. 2002; Birkofer 2007).
Measured parameters	The velocity profile, based on the signal frequencies reflected from inhomogeneous portions of flow.
An ultrasonic measurement system and a procedure	
Ultrasonic and signal processing techniques	Doppler UVP ultrasonic flowmeters. The PE technique (the TT technique is also possible). The signal cross-correlation technique (Theme 9).
Compatibility with other techniques	High-speed video and laser techniques.
No.	**7**
Study themes	Velocity profile (Kobayashi et al. 2006).
Measured parameters	Applicable for measuring the distribution profile of solids in liquid-particle flow. For example, a linear transducer array emits a plane wave into a slurry system and then the same transducer array records the backscattered waves, caused by a cloud of particles in the slurry.
An ultrasonic measurement system and a procedure	A linear transducer array or a single transducer. The transducer is mounted perpendicular to the flow.
Ultrasonic and signal processing techniques	The ultrasonic speckle correlation velocimetry (USV).
Compatibility with other techniques	High-speed video and laser techniques.

(continued)

Table 2.1 (continued)

No.	**8**
Study themes	Flow velocity (Endress+Hauser 2006).
Measured parameters	The average velocity determined from the measured time delay by ultrasounds. The relations between the sound speed and flow speed are expressed by: $c' = u + c$ $c'' = -u + c$ where: c—the sound speed in a medium u—the average flow speed of a medium
An ultrasonic measurement system and a procedure	A loop with a horizontal pipe section. Two transducers transmitting and receiving signals at the same time, located at two opposite sides of the horizontal section.
Ultrasonic and signal processing techniques	The transit-time difference (TTD) method. The PE technique.
Compatibility with other techniques	High-speed video and laser techniques.
No.	**9**
Study themes	Flow velocity (Time 1993).
Measured parameters	An analysis of small portions of a signal, searching for the time delays of the characteristic features of the signal.
An ultrasonic measurement system and a procedure	The average medium velocity is given by the following cross-correlation function (CCF): $\mathrm{CCF}(\tau) = \frac{1}{T}\int_0^T x(t)y(t+\tau)dt$
Ultrasonic and signal processing techniques	The cross-correlation of the ultrasonic echoes backscattered from particles. The PE technique.
Compatibility with other techniques	High-speed video and laser techniques.
No.	**10**
Study themes	The shape, surface reflectivity and location determination, based on the automated scanning of a solid body with carvings (Beckmann et al. 2006).
Measured parameters	The shape of a surface and the position of an object using the time delay measurement with the D-scan. The surface reflectivity by the amplitude magnitude measurement with the C-scan.
An ultrasonic measurement system and a procedure	The possibility of designing a pulse using a programmable signal generator (designing a short ultrasonic pulse). Using air-coupled ultrasounds, however, transducers must contain composite materials for matching the acoustical impedance (Bhardwaj 2000).
Ultrasonic and signal processing techniques	The time-domain spectroscopy. The ultrasonic imaging (the PE and TT techniques) and the automated scanning. Phase array transducers or a moving transducer.
Compatibility with other techniques	

Table 2.1 (continued)

No.	**11**
Study themes	Particle size determination using Rayleigh (surface) waves (Maharidge 1985).
Measured parameters	The sound amplitude measurement of the backscattered pulses (Rayleigh scattering) for calculating the size distribution of suspended particles in fluid. Low concentrations of particles; the sizes of 2–10 μm in water are applicable.
An ultrasonic measurement system and a procedure	A horizontal flow loop. One ultrasonic transducer.
Ultrasonic and signal processing techniques	Rayleigh backscattering of ultrasonic waves. The PE technique.
Compatibility with other techniques	
No.	**12**
Study themes	The inspection of certain geometries, such as flaws (Meyer and Rose 1997).
Measured parameters	Interrogating large areas of materials without extensive transducer scanning (perhaps from a single location). An appropriate mode must be selected for testing.
An ultrasonic measurement system and a procedure	The utilization of relatively broad-bandwidth transducers. The piezo-composite transducer design is applicable. The method's limitations are the following: – The method is not practical for materials thicker than 0.25 in. (0.6 cm). – It requires a basic understanding of the flaw geometry.
Ultrasonic and signal processing techniques	The guided wave inspection.
Compatibility with other techniques	

Table 2.2 Categories of ultrasonic measurements and their applications (Bhardwaj 2000).

Category	Measured parameters	Applications
Time domain	The times-of-flight and velocities of longitudinal, shear, and surface waves.	Density, thickness, defects detection, elastic and mechanical properties, anisotropy measurement. An interface analysis, proximity and dimensional analyses. Robotics and remote sensing applications.
Attenuation domain	The amplitude fluctuations at given frequencies in the reflected and transmitted signals.	The characterization of defects, surface and internal microstructures. An interface analysis.
Frequency domain	The sound frequency dependence using the ultrasonic attenuation spectroscopy method.	The determination of a microstructure, grain size, grain boundary relationships, porosity, the characterization of a surface.
Image domain	The time-of-flight, velocity and attenuation mapping by the raster C-scanning or the synthetic aperture techniques.	The surface and internal imaging of defects and microstructures. The density, velocity, mechanical properties measurement. 2D and 3D imaging.

Table 2.3 Types of ultrasonic transducers.

Transducer type	Description
A flat transducer	The most commonly used transducer with evenly distributed acoustic energy along the beam's path in the far-field area.
A focused transducer (spherical, cylindrical)	A focused transducer creates a narrow beam with high sound intensity, resulting in the improved resolution and sensitivity (Bai et al. 2024). It is used for microscanning, the examination of the smallest structures, due to its high frequency. The frequency range is up to 225 MHz (EVIDENT 2005). It is like an acoustic microscope. The transducer has designed focal volume. It requires long scanning times.
A transducer array	Multiple transducer elements are used in a single probe in a phased array transducer system. The elements are steered and focused at different angles and depths electronically, with the controlled timing of each element (NDT Products 2024). The electronic scanning (along the length of the linear array probe) creates the cross-sectional profile without moving the transducer. Two-dimensional transducer arrays are used for the medical diagnostic ultrasonic imaging.
A multifrequency transducer	Transducers work on two or more frequencies at the same time.

Table 2.4 Types of ultrasonic transducer couplings.

Type	Purpose	Comment
Contact	Direct-contact inspections	Using gel to exclude air ensures the proper transducer coupling to a tested material.
Non-contact (air-coupled)	Non-contact inspections	The acoustic impedance matching to air is considered in the design.
Immersion (water-coupled)	Underwater inspection	The acoustic impedance matching to water is considered in the design.

combined with the amplification of chosen portions of the reflected signal may bring promising results for measuring the particle concentration distribution.

The available pulser-receiver allows for the pulse-echo mode only and using a default spike pulse. Therefore, an external signal generator needs to be utilized for testing using different waveforms. The waveform generator is described in Sect. 2.2.2.

For the determination of average flow speed, two methods will be considered: the cross-correlation of backscattered signals and the transit-time difference methods (Themes 8 and 9 in Table 2.1).

2.2 The Improved Pulsed Broadband Ultrasonic Attenuation Spectroscopy Basics

The motivation of the project is to elaborate the methodology having potential for the online measurement of liquid-particle flow systems. This is fulfilled by the development of the improved pulsed broadband ultrasonic attenuation spectroscopy.

2.2.1 Selecting the Ultrasonic Transducer

In order to select the most optimal transducer for testing, which is sensitive and has high energy, three ultrasonic transducers from Panametrics of the natural frequencies of 2.25, 5, and 20 MHz and the corresponding sizes of 0.5, 0.25, and 0.125 in. were tested. The transducers have flat surfaces and are designed for the contact inspection. Figure 2.1a–c presents the spike signals with the 500 Hz pulse repetition frequency (PRF), reflected from a plastic wall, acquired in the pulse-echo mode.

Observations and conclusions are the following:

- Signal attenuation is greater for the 20 MHz transducer than for the lower frequency transducers (2.25, 5 MHz); therefore, it is better to use the latter ones, if signal amplitude needs to be analyzed.
- The low frequency transducers (2.25 and 5 MHz) are more sensitive in terms of their signal frequency spectra than the high frequency transducer (20 MHz), however, they produce a less stable signal than the latter one.
- The 5 MHz transducer performance seems to be the most optimal and it will be used for testing of liquid-particle flow.
- Further tests using different pulse shapes of various PRFs have confirmed that the 5 MHz transducer performance is very good for the testing purpose.

Other premises regarding the selection of the ultrasonic transducer for testing are the following:

- It is the most optimal to use two transducers of the same frequency for testing in the through-transmission mode.

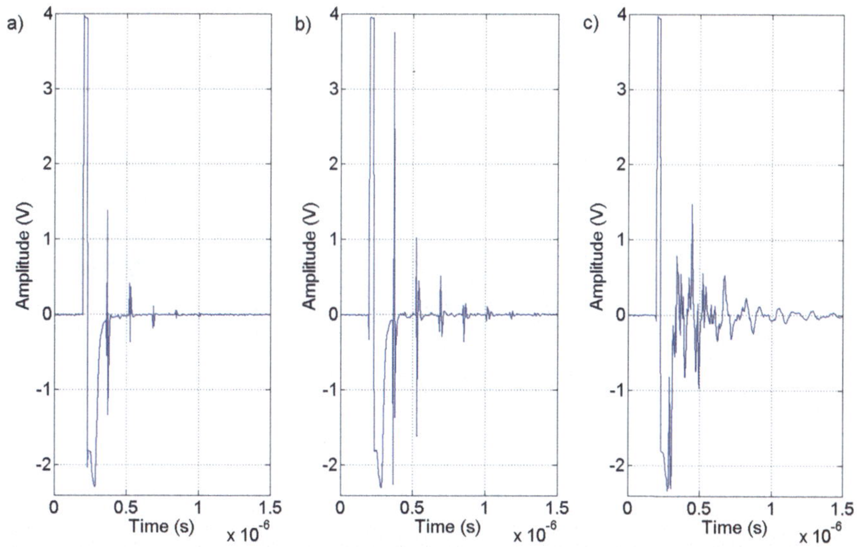

Fig. 2.1 The echoes from a plastic wall measured with the (**a**) 20 MHz, (**b**) 5 MHz, (**c**) 2.25 MHz transducers, using the 500 Hz PRF spike pulse.

- In order to detect particles of small size, it is recommended to use a transducer of the natural frequency providing a short enough wavelength ($\lambda = c/f$ and $f = 1/T$) compared to the particle size (the minimum particle size can be assumed as $\phi = \lambda/4$).
- The size of a transducer needs to be not too large compared to the size of particles.

2.2.2 *Experimental Setups*

Three types of mixing systems were used during the project work. A simple square cell of the size of 10 × 10 × 9 cm was used together with a magnetic stirrer placed on the cell's bottom, Fig. 2.2. The stirrer linear velocities in water were low (0.22–0.34 m/s), corresponding to the Reynolds numbers of 18,000–28,000. The flow was homogenous along the ultrasonic path. Ultrasound instrumentation components are the following, Fig. 2.2:

- The 33220A 20 MHz Function/Arbitrary Waveform Generator from Agilent, which generates electronic waves.
- Two standard-contact 5 MHz ultrasonic transducers of the size of 0.25 in. from Panametrics, which generate mechanical ultrasonic waves and receive them from a measured system.
- An ultrasonic preamplifier improving signal-to-noise ratios (SNR), powered by a direct current (DC) source.
- A computer unit with the installed NI PXI/PCI-5124, 12-Bit, 200 MS/s data acquisition card for digital signal acquisition.

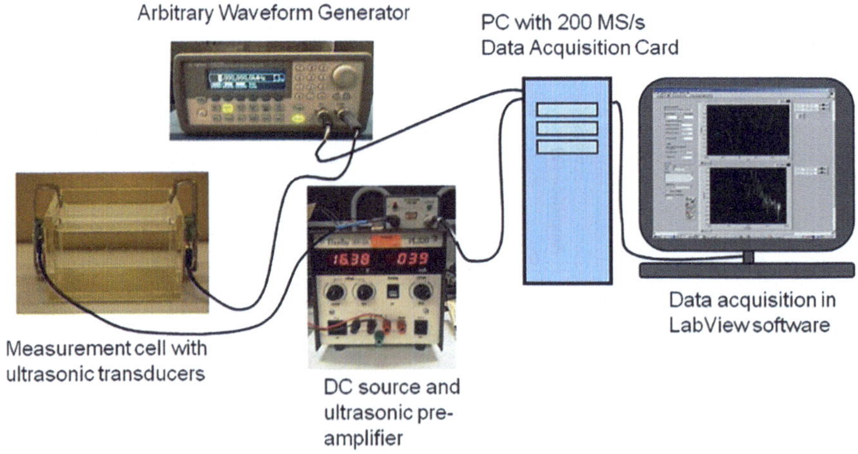

Fig. 2.2 The ultrasonic experimental setup for measurement of liquid-particle flow. (B.M. Wrobel, R.W. Time, *Applied Acoustics* 72 (2011) 324–335).

- The LabVIEW program displays a signal on a PC screen and saves it into a file containing time and amplitude data.

Studies of concentrated liquid-particle flow in high shear and turbulence were performed using a system containing a high-shear mixer, Fig. 2.3. The mixer velocities were ranging from 1.56 to 1.98 m/s and the Reynolds numbers were 39,600–50,500.

A rotational small-scale test cell is presented in Fig. 2.4. It has the 10 cm width, 9 cm height, and 157 cm mid-circumference. It was designed for more advanced studies of liquid and liquid-particle flows. A detailed design can be found in Sect. 3.2. The rotor velocities range from 0.01 to 3.5 m/s with the corresponding rotor's Reynolds numbers of 450–175,500. A rotating cover of the cell drags fluid around the annular flow channel, thus it creates the simple Couette flow of a constant shear rate from the top to bottom of the channel. The cell offers various possibilities of flow patterns (regimes), including particle suspensions along the cell's height, a particle bed and dunes forming on the cell's bottom. The formation of such flow structures depends on the properties of flowing media and flow velocities. Parameters possible for measurements are, for example, liquid and particle velocities, particle concentration, particle bed height, flow trajectories, and the rheological parameters of fluids. A velocity distribution profile along the cell's height (the axial velocity profile) and a concentration distribution profile across the flow channel can be measured. It is possible to perform flow velocity measurement using the transit-time difference method, and concentration measurement using the time-domain spectroscopy method. A bed height can be measured using the pulse-echo technique. The particle image velocimetry method using a laser and a high-speed camera can be used for validation of the transit-time difference flow velocity measurement results. The cell was used in the feasibility study of the ultrasonic transit-time difference flow velocity measurement (Sect. 3.3).

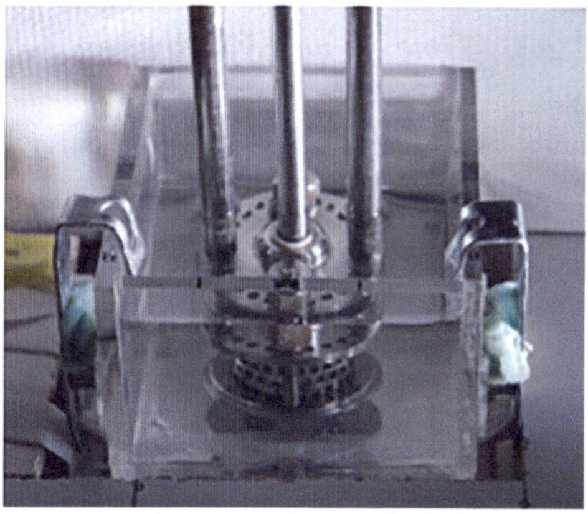

Fig. 2.3 The test cell with the high-shear mixer for ultrasonic measurement of liquid-particle flow.

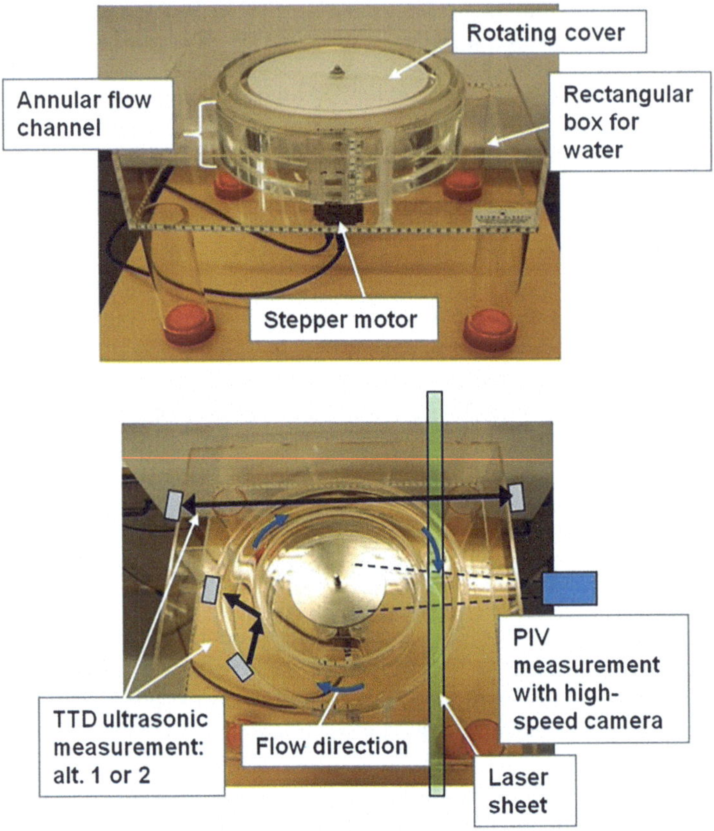

Fig. 2.4 The small-scale rotational flow cell for the ultrasonic transit-time difference and the particle image velocimetry measurements of liquid-particle flow.

2.2.3 Time Signals Verification

In the case when an acquired ultrasound signal needs to be verified, a good way of finding it and checking its quality is to compare this signal on both a computer screen and an oscilloscope, Fig. 2.5. Acquisition parameters can be adjusted in the LabVIEW program on the PC, based on the oscilloscope's settings. This method is very useful, when searching for a time-dependent signal coming at an initially unknown time.

2.2.4 Ultrasonic Waves Data Science

A high-quality ultrasonic wave is vital for high-accuracy testing, due to information carried in the signal. Proper design and acquisition of a wave enable high-quality information in the wave's clear and readable frequency spectra.

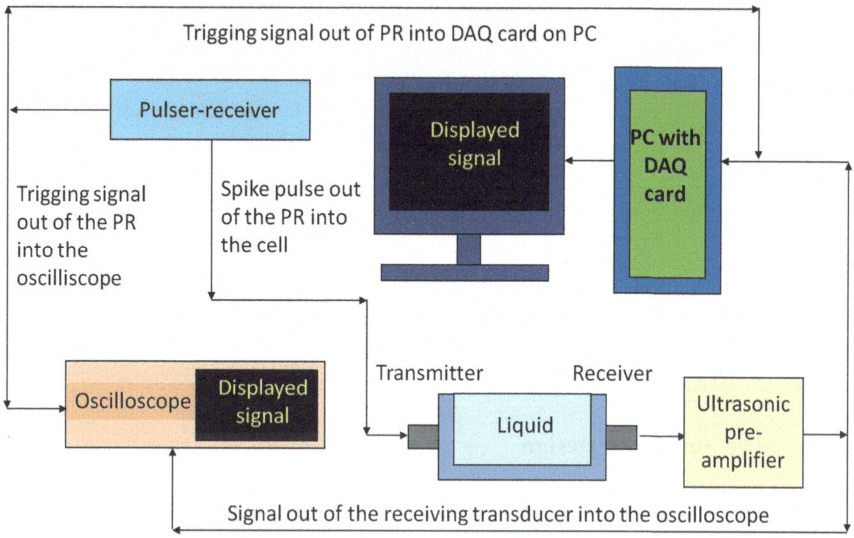

Fig. 2.5 A verification method for a time-dependent signal displayed on a computer screen and an oscilloscope.

2.2.4.1 The Ultrasonic Wave Design

The signal generator can output various waves of different shapes, adjustable pulse repetition frequency (PRF) and voltage, Fig. 2.2. However, the quality of the wave (the wave resolution) is default. Different methods for designing an external wave were studied. Such studies allowed answering a question regarding the influence of the wave resolution on the quality of its produced frequency spectrum. The experimental setup consisted of the waveform generator, the data acquisition card, the LabVIEW data acquisition program, and IntuiLink Waveform Editor with Visual Basic for designing waveforms.

A digital wave can be designed in various ways. The easiest way is to change wave parameters directly in the waveform generator; however, the maximum pulse repetition frequency is limited to 20 MHz and the number of points in a wave cycle (points per cycle) is fixed, which prevents the controlled acquisition of data points. An alternative is to use software, which is compatible with the generator to design a wave. While the IntuiLink editor allows for setting any number of points in a waveform, only some wave shapes are available. Writing macros in Visual Basic enables designing any wave shape with any number of points; therefore, this method was concluded to be the most beneficial. There is the 6 MHz limit for the maximum repetition frequency in IntuiLink. However, it is possible to design a wave containing many cycles and therefore obtain higher PRF within only one waveform.

IntuiLink Waveform Editor offers two designing modes: the ASCII Arb and the BINARY Arb. A wave designed in the BINARY Arb mode is sent to the waveform generator much faster than in the case of the ASCII Arb mode. Each point of the

Table 2.5 The limiting conditions for the wave design and acquisition with the chosen setup.

Hardware	Software	Operating ranges for chosen hardware models
Waveform generator	Visual Basic for wave design.	PRF: 1 μHz–6 MHz. Amplitude: 10 Vpp for the BINARY Arb mode and 5 Vpp for the ASCII Arb mode.
Data acquisition card	LabVIEW for data acquisition.	The maximum acquisition sampling frequency of the card is 200 MS/s.

wave can be either calculated using a wave equation or via assignation of a single value to it. All the points are gathered into an array and transferred to the wave generator, which outputs the designed wave (Table 2.5).

2.2.4.2 The Sine Wave Design

The design of the wave consisting of several cycles, with the wave frequency higher than the wave repetition frequency ($f_{\text{int}} > f_{\text{output}}$) is performed by the following method:

$n_{\text{cyc}} = 10$ (the number of cycles)
$p_{\text{cyc}} = 20$ (the number of points per cycle)
$p_{\text{wave}} = p_{\text{cyc}} \cdot n_{\text{cyc}} = 200$ (the number of points per wave)
$A = 10$ Vpp (the amplitude in peak-to-peak voltage)
$f_{\text{wave}} = 1$ MHz (the wave frequency)
$f_{\text{rep}} = f_{\text{wave}}/n_{\text{cyc}} = 100$ kHz (the repetition frequency or the output wave frequency)
$dt_{\text{wave}} = 1/f_{\text{rep}} = 1 \times 10^{-5}$ s (the delta time of the wave or the waveform period or the waveform duration)
$dt_{\text{point}} = dt_{\text{wave}}/p_{\text{wave}} = 5 \times 10^{-8}$ s (the delta time per point)
$f_{\text{sampling}} = 1/dt_{\text{point}} = p_{\text{wave}} \cdot f_{\text{rep}} = 20$ MS/s (the sampling frequency)
$f_{\text{sampling - req}} = 2 \cdot f_{\text{sampling}} = 40$ MS/s (the required sampling frequency)

2.2.4.3 The Wave Acquisition Principles

The basic rules for signal acquisition in terms of correctly calculated signal frequency by the fast Fourier transform (FFT) algorithm are presented below.

The accuracy of the wave frequency, calculated by the FFT depends on:

- An acquired full wave cycle (in the case of an incomplete wave cycle the accuracy is low).
- The number of points per cycle (the wave resolution)—the greater the wave resolution, the better accuracy of the calculated frequencies.
- The number of cycles per wave—generally, the higher number of cycles, the better accuracy of the calculated wave frequencies.

The accuracy does not depend on the wave amplitude.

The FFT amplitude of the frequency peaks depends on:

- The number of points per cycle. The higher the wave resolution, the higher amplitude. This dependency has a linear character.
- The number of cycles per wave. Theoretically, it does not influence the amplitude of the frequency peaks, which was confirmed by the simulation result (Sect. 2.2.9); however, in practical applications it is recommended to acquire at least a few wave cycles to provide better statistics for averaging.
- The wave amplitude—this dependency has an exponential character.

An advice is to acquire a wave with the sampling rate that at least doubles the wave resolution (referring to the Nyquist sampling rate theory).

2.2.4.4 The Nyquist Sampling Rate

For lossless digitization, the sampling rate should be at least twice the maximum frequency response. In principle, this should allow for the acquisition of all the wave points without losing them. The rule of thumb is that the higher sampling rate, the better.

2.2.4.5 The Sampling Rate Study

In order to obtain the high-quality and short ultrasonic pulses, with the pulse repetition frequency within the MHz range and containing the satisfying number of points per cycle (at least 100), a data acquisition card with a high sampling rate needs to be used.

The 200 MS/s data acquisition card provides the resolution of 5×10^{-9} s, which is high enough for the testing purposes (Figs. 2.6 and 2.7).

2.2.4.6 Signal Triggering

The signal generator in addition to continuously outputting waves used for testing, also sends out a synchronizing square pulse into the data acquisition card with a message to start acquisition. Triggering can start, for example, when the voltage change occurs, as it can be seen on the left side of Fig. 2.8. It is best to acquire a waveform only within one triggering interval, which is related to the repetition frequency, and repeat the acquisition several times for multiple measurements. This is due to possible wave continuity issues at the time of triggering, Fig. 2.9.

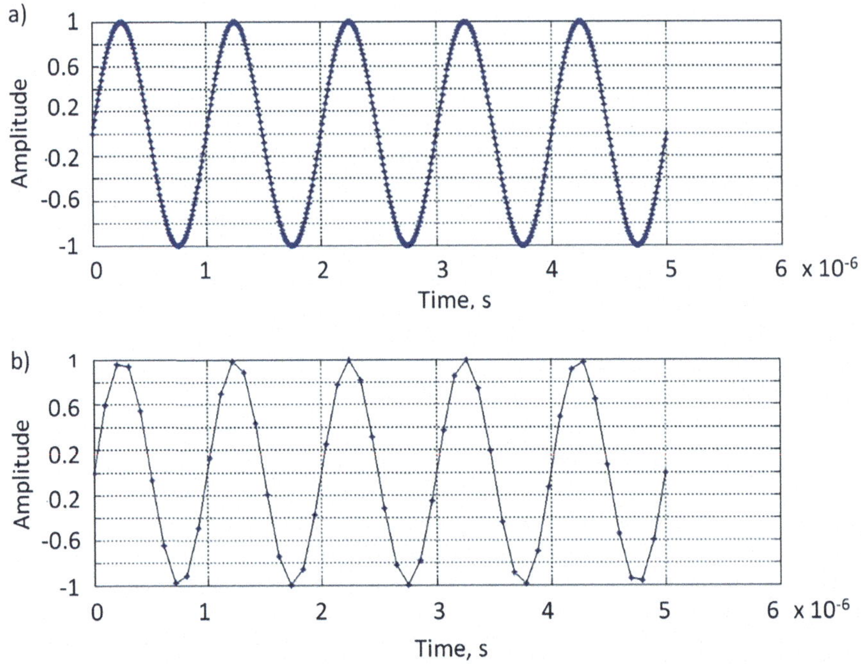

Fig. 2.6 The sine waves of (**a**) the high resolution (100 points per cycle) and (**b**) the poor resolution (10 points per cycle). The designed wave frequency is 1 MHz, while the calculated frequencies are 0.998 MHz (for the case a) and 0.98 MHz (for the case b).

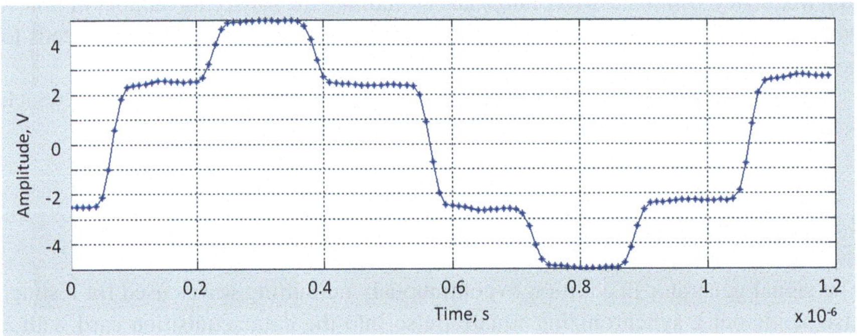

Fig. 2.7 The signal generator's response for the sine wave designed with the insufficient number of 6 points per cycle. The 100 MS/s sampling rate is much higher than it is required (12 MS/s).

2.2.5 Instrumentation Testing

2.2.5.1 The Signal Generator Testing

The signal generator was tested in order to determine the quality of driving signals. For this purpose, the sine waves of the frequencies from 100 kHz to 20 MHz and the

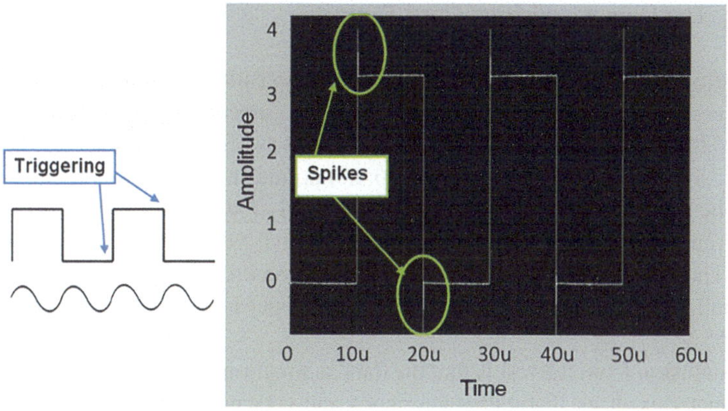

Fig. 2.8 On the LHS: triggering of a sine signal with a synchronizing square wave. On the RHS: the acquired triggering signal containing spikes.

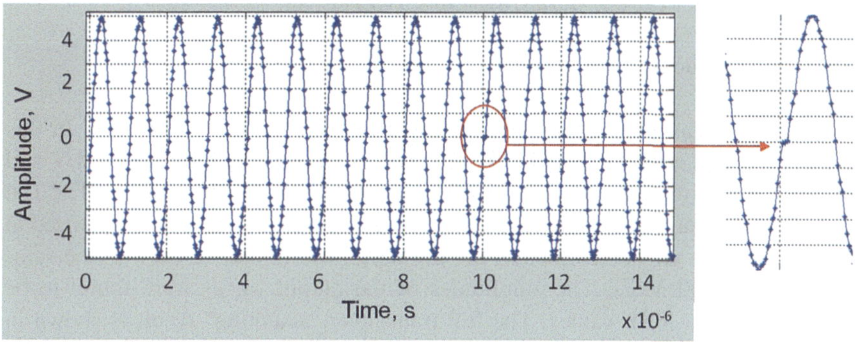

Fig. 2.9 The influence of the waveform triggering on the signal quality. The triggering point is after ten full sine wave cycles.

amplitude of 5 V were used. The acquisition sampling rate was 200 MS/s for all the tested waves. A deviation from the specified amplitude occurred for the 15 and 20 MHz waves, but there was no deviation for the 100 kHz–10 MHz waves. This may be associated with reaching the upper frequency limit of the waveform generator (corresponding to the smaller number of points per generated wave cycle, for the high PRF waves). An illustrating example would be: there are 2,000 points per cycle at 100 kHz, but this number is reduced to only 10 points per cycle at 20 MHz, to retain the same sampling rate. Otherwise, the 100 kHz–10 MHz waves are smooth, and the voltage is constant over the whole time range.

A conclusion from the test is that the high-quality waves up to 15 MHz can be obtained using the signal generator and 5 MHz ultrasonic transducer.

2.2.5.2 The Data Acquisition Card Testing

The logging data acquisition card was tested to determine the sensitivity and linearity of the direct current (DC) levels. The direct current source was directly connected to the data acquisition card and the signal was acquired on the computer. The acquired signal in the test consisted of 20,000 points and the sampling rate was 200 MS/s. The tested voltage ranges were 0.03–1 and 1–7 V. The measured mean versus the reference voltages were acquired. The residual voltage varied from 0.003 to 0.04 V. The calculated standard deviations of the recorded continuous voltage were very small from 3×10^{-12} to 4.4×10^{-3} V. The upper logging card limit was found to be 5.65 V.

A conclusion from the test is that the data acquisition card showed good sensitivity associated with its 12-bit characteristics and very good linearity tested with the DC voltages.

2.2.5.3 Testing of the 5 MHz Ultrasonic Transducer

The Single Frequency Sine Waves

The 5 MHz ultrasonic transducer was tested to determine its bandwidth. In this test, the transducer was driven by the sine waves with the amplitude of 2.5 V and the pulse repetition frequency (PRF) values ranging from 100 kHz to 20 MHz. The waveform generator and the square test cell that was filled with water were used in the test. The test was performed in the through-transmission mode with the sampling rate of 200 MS/s. The amplitudes of the output waves were found to be dependent on the PRF values. The full transducer "scanning" result is shown in Fig. 2.10. The maximum acoustic power out of any ultrasonic transducer is close to its resonance frequency. The signal amplitudes larger than 0.2 V were obtained for the PRF values from 500 kHz to 10 MHz. This is related to the transducer's frequency useful for testing. However, the response curve deviates from what is expected for the ultrasonic transducer, due the considerable scattering of the data points.

The scattering effect was investigated in detail in repeated tests. They are shown in the inserts in Fig. 2.10. It turned out that the cell's walls reflected the waves back and forth, causing internal resonances ("the ringing"). When the pulse repetition frequency parameter was varied in small steps, the interference conditions were repeating approximately every 8 kHz for the wall-wall reflections for this specific setup. The wavy signal's patterns of the sound interference need to be accounted for in practical applications. Parameters that influence the frequency of wave interference are the cell's dimension and the sound speed in fluid. The wave interference effect is discussed in more detail in Sect. 2.2.6.

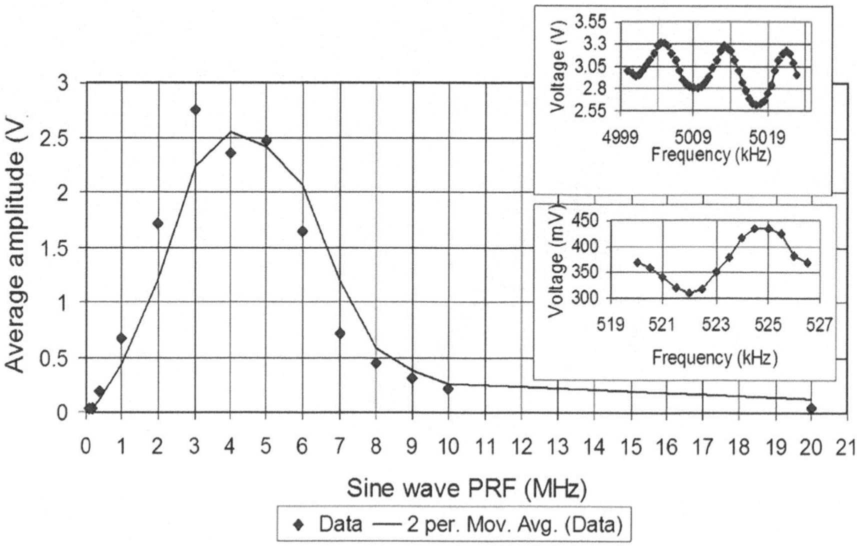

Fig. 2.10 The frequency response of the 5 MHz transducer, based on scanning with the 100 kHz–20 MHz, 2.5 V amplitude sine waves. The figure inserts show magnified views, illustrating the sound interference. (B.M. Wrobel, R.W. Time, *Applied Acoustics* 72 (2011) 324–335).

The Series of Sine Sweeps

In order to verify the above described test, in which individual frequency waves were used, the series of sine sweeps from 500 kHz to 20 MHz were used this time (500 kHz–20 MHz, 1–20 MHz, 1.5–20 MHz, …., 19–20 MHz), Fig. 2.11. The frequency division is dictated by a limited amount of points for signal acquisition, due the computer memory and long acquisition time. The measurement was performed using the square-shaped cell filled with water (at static conditions) and two 5 MHz ultrasonic transducers set up in the through-transmission mode. "Only" the first three million points of the signal could be acquired simultaneously (each preceding sweep series overlaps with the next one). It can be observed that the shape of the frequency spectrum in Fig. 2.11 is similar to the one in Fig. 2.10. The noise starts around 12 MHz, where the amplitude is smaller than 0.0008 V and this can be seen in Fig. 2.12. Even the sweep 19–19.15 MHz is detected, however, its amplitude is buried in the noise level.

The results confirm the acquired earlier results from the test with individual frequency sine waves and indicate on the final useful frequency range for the 5 MHz ultrasonic transducer, which is from 500 kHz to 12 MHz.

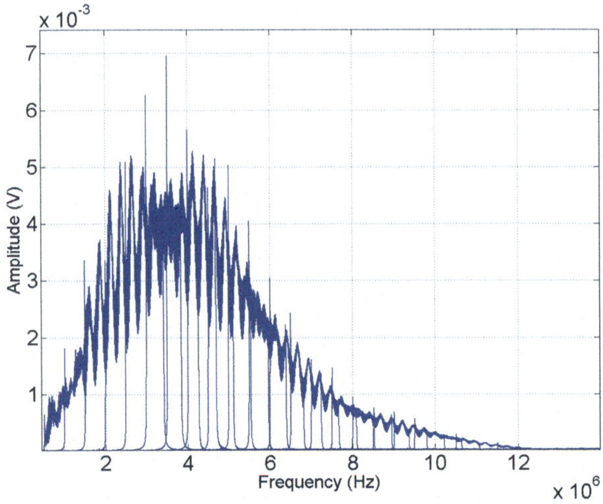

Fig. 2.11 The spectral amplitudes of the 500 kHz–14 MHz sine sweeps, obtained in the through-transmission test for water at static conditions, using the 5 MHz ultrasonic transducers.

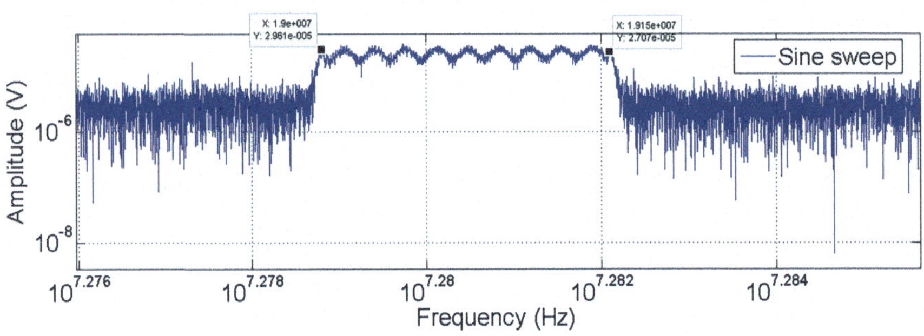

Fig. 2.12 The 19–19.15 MHz sine sweep buried in the signal's noise level.

2.2.6 Wave Interference

Wave interference occurred during the tests with the square-shaped cell filled with water, Fig. 2.13, and two types of incident waves: the 500 kHz–20 MHz linear sine sweep with the duration of 2 ms, Fig. 2.14a, and the 10 kHz square wave with the duration of 0.1 ms, Fig. 2.15a. The measurement was performed in the through-transmission mode. The wave interference effect is shown in the frequency spectra, Figs. 2.14b and 2.15b. This effect is represented by the waviness of sound amplitude plotted versus frequency.

The mechanism of the wave interference is based on the superposition of the reflected waves from walls with the direct source wave. A condition for the constructive interference is that after a certain number of wavelengths, the direct wave

Fig. 2.13 The generation of wave interference.

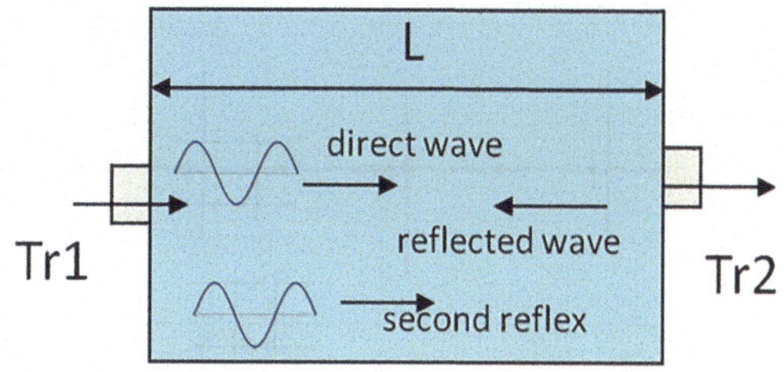

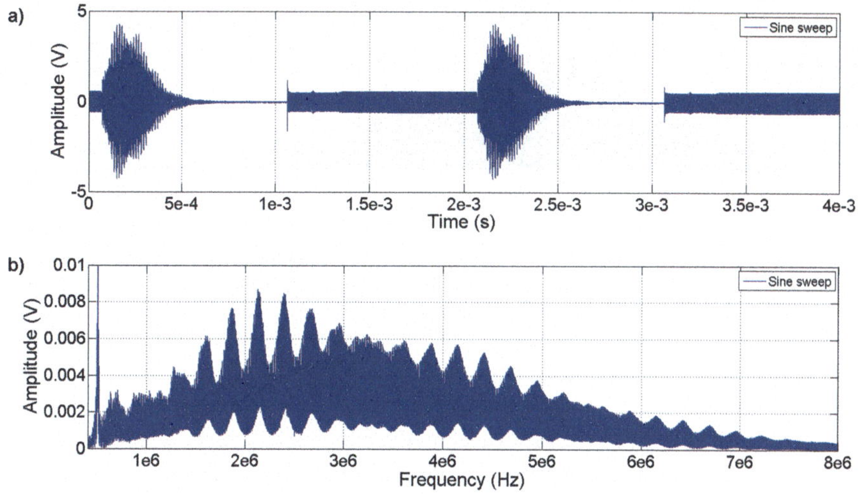

Fig. 2.14 (**a**) Two cycles of the 500 kHz–20 MHz linear sine sweep with the duration of 2 ms, obtained during the ultrasonic through-transmission test for the square-shaped cell filled with water. (**b**) The frequency spectrum is influenced by the wave interference.

and the reflected wave (after traveling through the distance L) are in phase: $n \cdot \lambda = 2L$. The wave frequency, f, and the wavelength, λ, are related through: $f \cdot \lambda = c$, where c is the sound speed in a tested medium.

2.2.6.1 An Example of the Wave Interference Calculation

Based on the test data in Fig. 2.10 (the small figure), the wave frequency increase from 5.2×10^5 to 5.27×10^5 Hz brings the amplitude back to the same level. Assuming that one full wavelength is required for the wave amplitude amplification ($n_2 = n_1 + 1$), what are the number of wavelengths and the wavelength, for the wave frequency increasing from 5.2×10^5 to 5.27×10^5 Hz, after the wave passes through the distance $2L$ in water?

$L = 0.1$ m (the distance)
$f_1 = 5.2 \times 10^5$ Hz (the initial wave frequency)

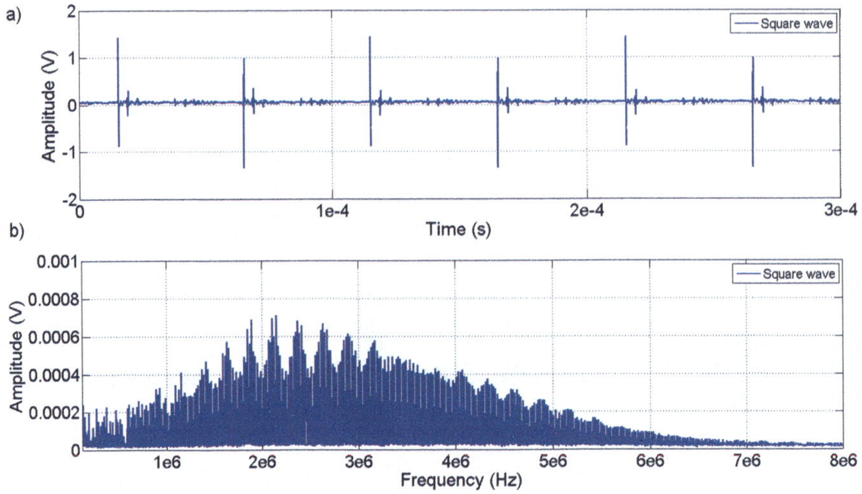

Fig. 2.15 (**a**) Three cycles of the 10 kHz square wave with the duration of 0.1 ms, obtained during the ultrasonic through-transmission test for the square-shaped cell filled with water. (**b**) The frequency spectrum is influenced by the wave interference.

c_{water} = 1500 m/s (the sound speed in water)
$\lambda_1 = c_{water}/f_1 = 2.88 \times 10^{-3}$ m (the wavelength in water)
$n_1 = 2L/\lambda_1 = 69.33$ (the number of wavelengths within the distance $2L$)
$n_2 = n_1 + 1 = 70.33$ (the next number of wavelengths)
$n_2 \cdot \lambda_2 = 2L$
$\lambda_2 = 2L/n_2 = 2.84 \times 10^{-3}$ m (the next wavelength)
$f_2 = c_{water}/\lambda_2 = 5.275 \times 10^5$ Hz (the new frequency)

2.2.7 *A Square Wave as the Novel Broadband Pulse*

A square wave is built from many sine functions (Fig. 2.16). It contains naturally the fundamental frequency and the harmonic frequencies. There are two square wave equations available, based on the sine function (2.1) and the signum function (2.2):

$$x_{\text{square}}(t) = \frac{4}{\pi}\sum_{k=1}^{\infty}\frac{\sin\left((2k-1)2\pi ft\right)}{2k-1} = \frac{4}{\pi}\begin{pmatrix}\sin(2\pi ft)+\frac{1}{3}\sin(6\pi ft)\\ +\frac{1}{5}\sin(10\pi ft)+\cdots\end{pmatrix} \tag{2.1}$$

$$x_{\text{square}}(t) = A\cdot sign\left(\sin(2\pi ft)\right) \tag{2.2}$$

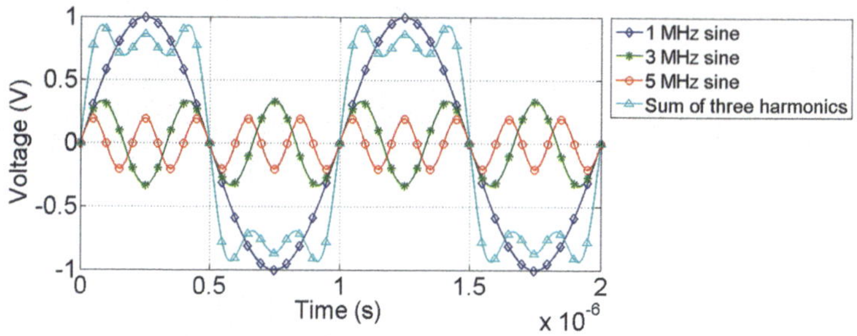

Fig. 2.16 The 1 MHz square wave built from three sine waves.

The square wave is mathematically equivalent to the sum of the sine waves having the fundamental frequency, plus an infinite series of the odd-multiple frequencies of diminishing amplitudes (the harmonics). The frequencies of the harmonics are multiples of the fundamental wave frequency. The frequency of each sine wave that builds the square wave can be found by processing of the signal by the fast Fourier transform. The harmonics have the amplitude of the size of $1/n$, relative to the amplitude size of the fundamental frequency sine wave.

The square wave is a novel broadband pulse shape investigated for the purpose of testing liquid-particle flow. There are a few advantages of using it in the ultrasonic attenuation spectroscopy, over the other types of broadband pulses, which have been traditionally used for testing, International Organization for Standardization (2006):

- A very short duration time (all the sine cycles are within one square cycle). Pulses such as linear sine sweeps and bursts, International Organization for Standardization (2006), last for a much longer time. One cycle of the square wave lasts for 0.1 ms and of the linear sweep for 2 ms, Figs. 2.14a and 2.15a in Sect. 2.2.6. Fast measurement is essential in the case of quickly changing dynamic flow systems.
- Square pulses are significantly better suited for the wave energy conversion by the ultrasonic transducer (without noticeable signal losses), than spike pulses (the alternative broadband pulses of short durations). The square pulse provides twice higher spectral energy than the spike pulse, Fig. 2.17. This is due to the electro-mechanical inertia, causing the damped transducer output response in the case of the short duration (50 ns) spike pulse, compared to the longer duration square pulse (0.1 ms), Fig. 2.18.

2.2.8 *The Fast Fourier Transform*

A fast Fourier transform (FFT) is a method of converting time-domain signals into frequency-domain signals. A frequency spectrum can be generated via the FFT of the signal. The output parameters are the amplitude and the phase, which depend on the frequency.

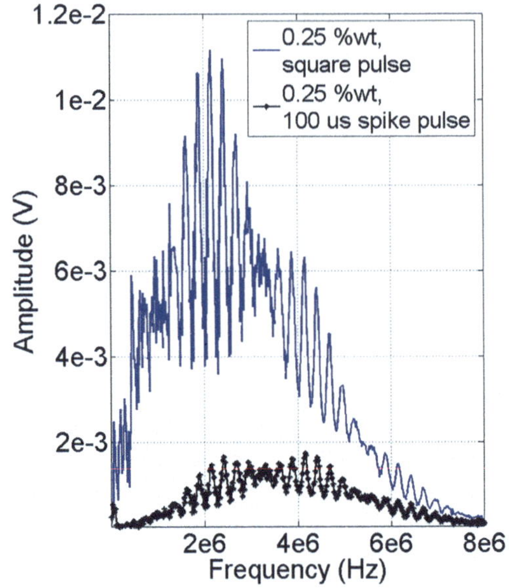

Fig. 2.17 The comparison of the frequency spectra for the 0.25$\%_{wt}$ water-glass particle suspension, using the 10 kHz square and spike pulses to drive the 5 MHz ultrasonic transducer. (B.M. Wrobel, R.W. Time, *Applied Acoustics* 72 (2011) 324–335).

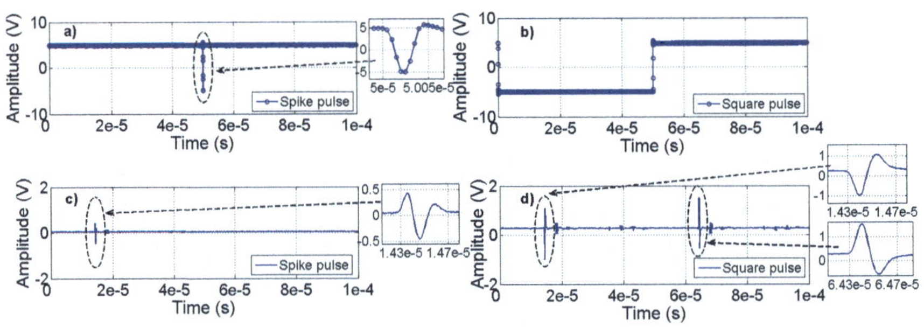

Fig. 2.18 The 10 kHz spike (**a**) and square (**b**) pulses acquired directly from the signal generator. The through-transmission test results for water acquired with the 5 MHz transducer, using the spike (**c**) and the square pulses (**d**). (B.M. Wrobel, R.W. Time, *Applied Acoustics* 72 (2011) 324–335).

The fast Fourier transform, $G(\omega)$, is the frequency function with real and imaginary parts, Eqs. (2.3) and (2.4). The parts correspond to the amplitude and the wave phase.

$$G(\omega) = \int_{-\infty}^{\infty} f(t) e^{-i\omega t} dt \tag{2.3}$$

$$G(\omega) = \int_{-\infty}^{\infty} f(t)\cos(\omega t)dt - \int_{-\infty}^{\infty} f(t)\sin(\omega t)dt \tag{2.4}$$

According to Euler's formula: $e^{i\omega t} = \cos(\omega t) + i\sin(\omega t)$, therefore, $e^{-i\omega t} = \cos(-\omega t) + i\sin(-\omega t) = \cos(\omega t) - i\sin(\omega t)$. The wave frequency, ω, is related to the wave period, T, and the number of cycles, n, by $\omega = 2\pi f = 2\pi(n/T)$.

A comment: to obtain the spectral amplitude in absolute values using an FFT algorithm, a custom function may need to be used.

2.2.9 *Numerical Simulations*

Numerical simulations were performed in MATLAB to visualize how an ultrasonic signal might be affected, after traveling through a medium, Fig. 2.19a–c. In Fig. 2.19a, the noise or the medium's response is represented by the sine wave containing both 10 and 40 MHz frequencies. The 20 MHz PRF square wave simulates the incident ultrasonic wave, Fig. 2.19b. The sum of these two waves represents the received signal, after passing through the medium, Fig. 2.19c. The simulations show that the medium's frequencies (the noise) could in theory be easily detected in the background of the incident wave's harmonic frequencies. However, as the real experiments have showed (Sects. 2.4 and 2.5), this mechanism does not apply. The number of incoming and outcoming frequencies is constant, and only the peak amplitudes of the fixed frequencies are really used for measurement, Fig. 2.19b.

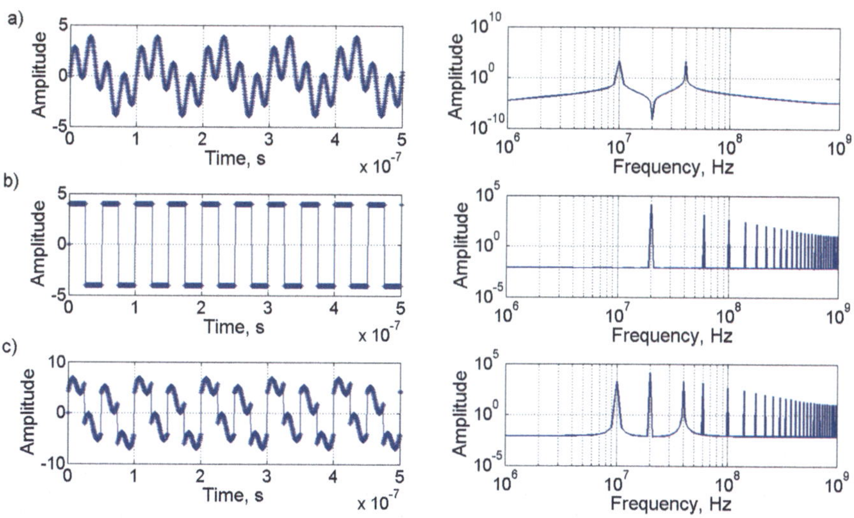

Fig. 2.19 The simulation results of (**a**) the sine wave consisting of 10 and 40 MHz frequencies, treated as the noise or the medium's response, (**b**) the 20 MHz square wave, treated as the incident ultrasonic signal for testing, (**c**) the square wave only theoretically affected by the noise. The corresponding frequency spectra are on the RHS.

Numerical simulations were also used for defining the wave acquisition principles (Sect. 2.2.4).

2.2.10 *Square Wave Design*

Studying different designs of square waves (shapes, various numbers of points in the waves' sections) resulted in a conclusion that the most optimal design providing high-spectral energy is of the waves which have:

- Short rise/fall times (related to the vertical parts).
- A high number of points in the rise/fall times parts.

The number of sine waves in the sum, Eq. (2.1), influences the lengths of the pulse rise/fall times. The higher this number is, the shorter rise/fall times become, resulting in the high-quality frequency spectrum. On the other hand, when vertical wave sections are well-defined (by many points), then the sine waves "in the sum" are well-defined also, resulting in the high-power spectrum. The amplitudes of the even-multiple frequency harmonics, which are normally much smaller than the amplitudes of the odd-multiple frequency harmonics, can be increased by applying an unsymmetrical square wave shape. This will result in the increased total resolution of the wave spectrum.

2.2.11 *Signal Averaging Data Science*

Noise occurs frequently during acoustic measurements. For example, the electronic noise in the radio frequency ranges from MHz to GHz, the flow induced noise, which is linked to dynamic nature of flow processes (kHz to several MHz), and the noise due to the variation of particle mass fraction along the acoustic beam, affecting high frequencies (1–100 MHz). To improve the quality of signals used for the ultrasonic attenuation spectroscopy (the signal-to-noise ratio and the method's reliability), it is recommended to use signal averaging techniques.

2.2.11.1 The Ensemble Average Technique

The ensemble average of signals is based on adding up individual, sequential frequency spectra and then dividing the outcome by the total number of samples. Figure 2.20 shows the result of the averaging of the simulated noisy square signals (a) and the measured signals (b), based on 30 spectral samples. The time signal consisted of 1,000 wave cycles and it was split into the smaller time intervals of 20 cycles, with the FFT applied on these individual sets. Then, the final averaged spectrum was calculated using the ensemble average technique (the corresponding

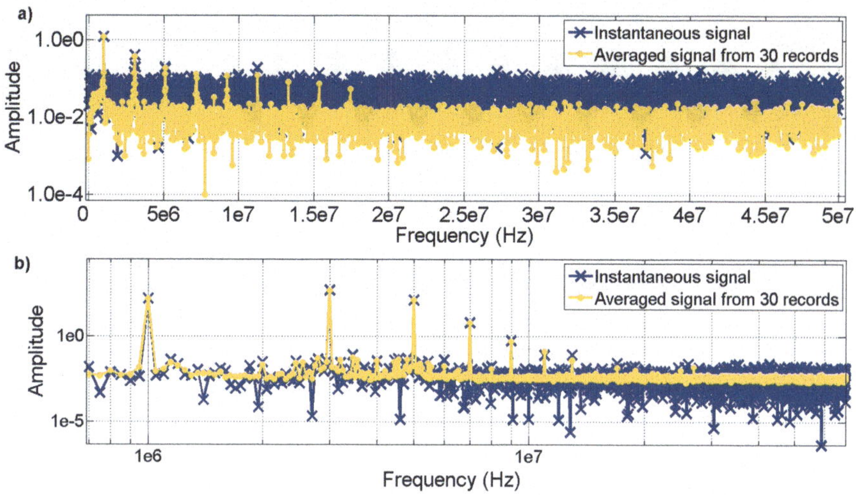

Fig. 2.20 The simulated (**a**) and the measured (**b**) frequency spectra of two signals: the instantaneous square signal and the ensemble averaged signal (based on 30 samples). The spectral peaks up to 17 MHz appear well above the noise level in the average case. (B.M. Wrobel, R.W. Time, *Applied Acoustics* 72 (2011) 324–335).

MATLAB code is in Sect. 5.1.3). The averaging of time signals, instead of their spectra, holds a risk of losing information at the time-domain stage (the modified time signal will result in the modified spectrum). Therefore, the averaging of "raw" spectral results is a better choice instead.

The advantages of the ensemble average method are the following:

- Improvement of the quality of a frequency spectrum; even the high-frequency peaks, which are normally buried in instantaneous signals, can be detected above the noise level.
- Suitability for the averaging of several frequency spectra.

The disadvantages of the ensemble average method are the following:

- It requires acquisition of multiple signal samples during a test, which may be time consuming.
- It is a less precise averaging method for time signals, compared to the running average technique.
- The quality of the averaging of challenging signals with high-frequency noise may not be satisfying, in terms of exposing the frequencies within the high range.

2.2.11.2 The Running Average Technique

The running average technique (or the moving average) was used for averaging signals with unacceptable amounts of random noise, like the one shown in Fig. 2.21a. The corresponding MATLAB code is in Sect. 5.1.4.

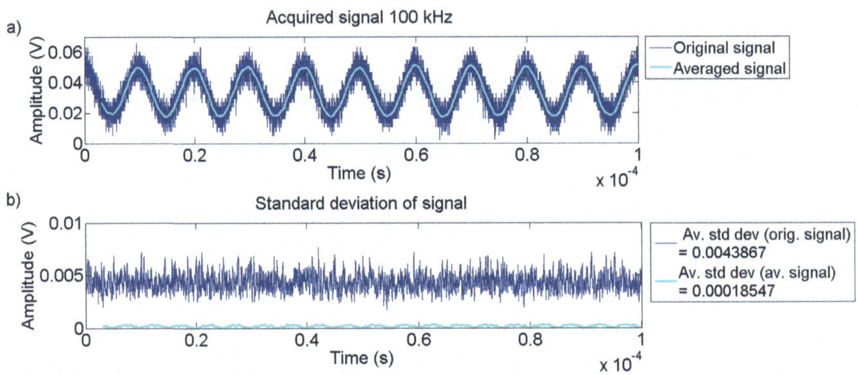

Fig. 2.21 (**a**) The acquired original and running averaged 100 kHz sine waves. (**b**) The instantaneous and average standard deviations of the original and average signals.

The methodology is based on the calculation of one average point from a narrow "averaging window" consisting of, for example, 20 consecutive data points (compared to 80,000 points acquired in a single test). Then, another average point is calculated from the set of the next 20 consecutive points, etc. The repeating scheme is that 19 points always overlap from the previous set and one point is newly added. The averaging is based on running several loops. The average amplitude points at the next iteration step, $n + 1$, are based on the average amplitude points calculated at the previous step, n; therefore, the matrix size consisting of the average amplitude (and the corresponding time points) decreases, for an increasing number of iterations. The instantaneous standard deviations of original and average signals are also calculated for the control, Fig. 2.21b. Similarly, as in the case of the amplitude points, this calculation is based on the sets of 20 consecutive points and is plotted in a time-domain graph. Additionally, the final average standard deviation corresponding to all data points is given for the quantification and control. The averaging procedure is repeated for a certain number of iterations until the break condition is fulfilled, which is the monotonicity of the selected part of the curve, for example, a few (5) consecutive points gradually increase (for the climbing curve's part) or decrease (for the falling curve's part). Then, a defined number of extra iteration loops ensures the curve's smoothness. The break condition is also triggered, when the number of data points in the original input file is too low to perform averaging.

The advantages of the running average method are the following:

- A short testing time—only one sample is required for averaging.
- Precise averaging of time signals.

When comparing the running average method to the FFT-based filtering (the ensemble average), the advantages are the following:

- More "controlled averaging"—the adjustment of a desired noise level and therefore better preservation of an original signal shape.

- The avoidance of the unacceptable phase shift and distortion of an original time signal, caused by the FFT filtering due to utilizing the cut-off frequency. In the case of reverting the processed signal by the FFT, from the frequency to time domain, the resulting signal will not be the same as the original time signal.

The disadvantages of the running average method are the following:

- Time consuming computing related to the narrow "averaging window" to get precise results.
- Losing the front and end parts of an original signal, due to averaging.
- When the break condition for iterations is not set correctly, there is a possibility of overaveraging leading to the amplitude reduction.
- Unsuitable for raw spectral signals (only applicable for time signals).

2.2.12 Acquired Signals Post-Processing

The post-processing of acquired signals was inherent to the project after the experimental testing. The averaging, transforming, and selective 3D plotting processes were computing memory-demanding, due to dealing with big datasets (the sets of 60,000–80,000 points or larger). Written programming codes were at a relatively advanced level. The selected programs are available in Chap. 5. Loops and subloops were utilized for automation. The codes contained a number of variable arrays, memory-clearing commands, "if" loops as the stop conditions, "while" conditions and "for" loops.

2.2.13 Conclusions

- The selected 5 MHz ultrasonic transducer (Sect. 2.2.1) is suitable for testing of liquid-particle flow. It has good sensitivity and it provides the output waves of high energy and good-quality pulses, within its intermediate frequency range of 500 Hz–20 MHz.
- The 200 MHz data acquisition card provides the resolution of 5×10^{-9} s, which is suitable for testing of the applications within the project scope.
- An incident wave for an ultrasonic transducer can be designed by a user using software compatible with a waveform generator. This method enables studying the wave's parameters (the sampling rate, the pulse repetition frequency, the amplitude) and shapes. It is a flexible method for constructing the wave with the parameters outside the default generator's limits. This method enables designing high-resolution waves.
- The triggering of a signal is important for the acquisition at the same fixed point. A square wave is used as the triggering pulse in the utilized wave generator. Triggering may introduce minor noise in a signal, because the triggering pulse is

not perfect, as it contains spikes. This can be avoided by using the high-quality triggering waves of a high number of points per cycle.

- Elements of any instrumentation contain their own measurement limits and will demonstrate specific behavior during testing; therefore, they need to be tested individually to find the common, optimal parameters, which ensure good accuracy testing.
- The wave interference effect (which can be seen in a signal as the "waviness," or "ringing") needs to be taken into consideration in practical applications.
- Any square wave provides a broadband frequency spectrum. Practically, the 1 MHz PRF square wave can provide the spectrum up to 100 MHz, when it is not damped by a medium. It can be used for testing using the pulsed broadband ultrasonic attenuation spectroscopy technique. It is better suited for energy conversion by the ultrasonic transducer than a spike pulse, which is also a short broadband pulse. It has the advantage of being short, compared to sweeps and bursts, which is essential for fast testing of dynamic systems.
- The square wave of short rise/fall times and a high number of points produces a high-quality spectrum.
- Signal acquisition needs to be planned to process a noisy signal using averaging techniques. The running average and the ensemble average techniques are useful to apply to ultrasonic signals for noise removal.
- The MATLAB simulations illustrated only theoretically the mechanism of ultrasonic testing of a medium using a square wave. The hypothesis was, whether it is possible to detect extra frequencies produced by the medium, on the background of the incident square wave's frequencies. The practical experimental results have showed (Sects. 2.4 and 2.5) that this mechanism does not apply. This is because the number of incoming and outcoming frequencies is constant and only the peak amplitudes of the incident wave's frequencies are essential for testing.
- The simulations were also used for defining the wave acquisition principles (Sect. 2.2.4).
- The post-processing of the ultrasonic data requires writing programs of an advanced level, which include computing memory-clearing commands, variable matrices, stop conditions, calculation loops and subloops.

2.3 Testing the Pulsed Broadband Ultrasonic Attenuation Spectroscopy

2.3.1 Subproject Objectives and Challenges

2.3.1.1 The Subproject Objectives

- The objective is to find the suitability of the pulsed broadband ultrasonic attenuation spectroscopy for testing of liquids having different physical properties.

2.3.1.2 The Subproject Challenges

- Finding the sensitivity of the spectroscopy method, in terms of measurement of liquids having different physical properties.
- Finding the sensitivity of the spectroscopy method to particle sizes.

2.3.2 Tests for Different Liquids at Static Conditions

According to the ISO standard for the ultrasonic attenuation spectroscopy from International Organization for Standardization (2006), the sound attenuation spectroscopy can be used for characterizing fluids having different physical properties, as well as for dispersed particles of various sizes. To verify this, a few liquid samples of different physical properties (glycerol, silicon oil, NaCl brine, and water) were tested. A square pulse was used as the incident ultrasonic signal. The measurements were performed in the through-transmission mode, using two 5 MHz transducers and the sampling rate was 200 MS/s. The liquid samples were measured at static conditions in the plastic square-shaped cell with the width of 10 cm, Fig. 2.2. The utilized square wave was designed in the ASCII Arb mode (Sect. 2.2.4) and generated in the signal generator, Fig. 2.2. The emitted wave traveled through the cell and the liquid, and it was received on the other side of the cell. The LabVIEW program acquired the data. Results are presented in Figs. 2.22 and 2.23 and the physical properties of the liquids are listed in Table 2.6. The original reference signal was acquired directly from the signal generator and it is presented for comparison in Figs. 2.22a and 2.23a.

The following observations were made, based on the tests:

- The amplitudes of the harmonic frequencies were attenuated by different rates in the tested media. This can also be seen as different signal shapes of the time signals acquired for individual liquids.
- The liquids of high viscosity (glycerol, silicon oil) dampened the ultrasonic signal to a greater degree, than of low viscosity (water), Figs. 2.22 and 2.23 and Table 2.6.
- The number of the incident wave's frequencies, consisting of the fundamental frequency (1 MHz) and its harmonics, remained unchanged in the output signals.
- The acquired reference square wave contains in addition to the odd frequencies, also the even ones of smaller amplitudes. This is a practical phenomenon, in the case of a measured square wave.
- The results are specific for the plastic square-shaped cell used in the experiment.

The colloid literature gives a lot of examples of the measurement of sound attenuation in media having different properties, through the frequency spectrum. For example, Dukhin et al. (2005) measured dairy products, which resulted in greater sound attenuation in high-fat milk, than in low-fat milk. In principle, this example confirms the results obtained for low and high viscosity liquids. Assuming that the

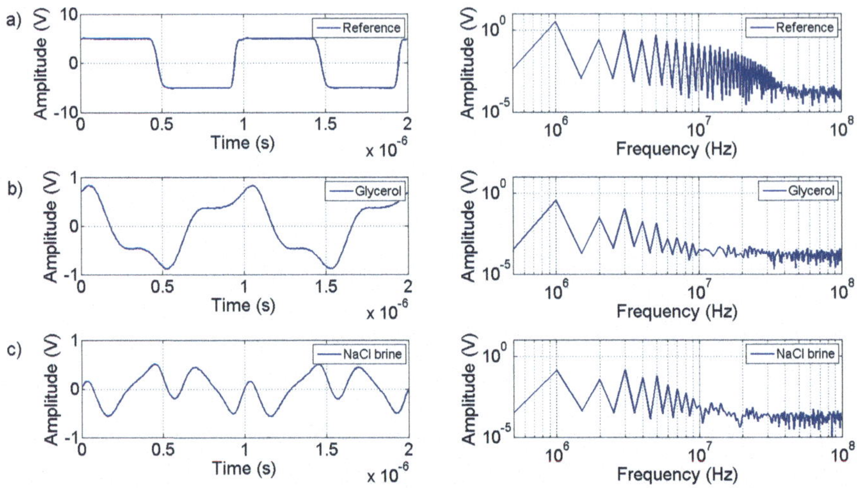

Fig. 2.22 (**a**) The reference 1 MHz square signal acquired from the signal generator with its spectrum. The measured signals and spectra were obtained for (**b**) glycerol, (**c**) NaCl brine.

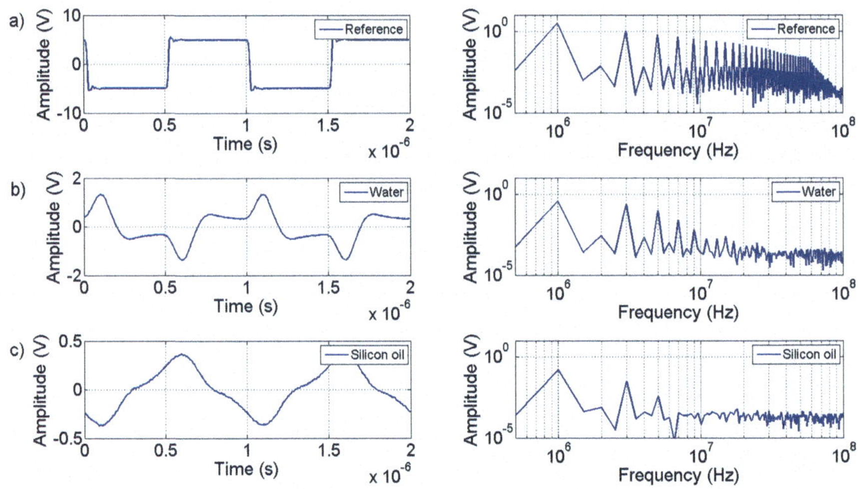

Fig. 2.23 (**a**) The reference 1 MHz square signal acquired from the signal generator with its spectrum. The measured signals and spectra were obtained for (**b**) water, (**c**) silicon oil.

sound loss is only due to viscosity and the other losses are negligible, which is generally true for liquids, according to the superposition theory (Sect. 2.5.3), the sound losses in water, glycerol, and silicon were calculated in Table 2.7, using Stokes' equation (Eq. 2.16, Sect. 2.5.4). The calculated values indicate that the viscous losses are the largest in glycerol and the smallest in water (independently on the sound frequency) and this validates and confirms the experimental results.

Table 2.6 The physical properties of liquids.

Parameter	Water	Glycerol	Silicon oil
Density (kg/m^3)	997	1230	930
Viscosity (Pa·s)	0.001	1.5	0.01055

Table 2.7 The calculated viscous sound attenuation in water, glycerol, and silicon oil using Stokes' equation (Eq. 2.16, Sect. 2.5.4).

Liquid	Viscous sound attenuation, α_s (Np/cm)	
	At 10 kHz	At 10 MHz
Water	7.8×10^{-9}	7.8×10^{-3}
Glycerol	4.4×10^{-6}	4.4×10^{0}
Silicon oil	1.2×10^{-7}	1.2×10^{-1}

It has been a challenge to find in the literature examples on studying broadband pulse shapes in their time domains. However, this feature of a signal seems to be quite valuable since it is like a "signature" or a pattern, which is specific for individual liquids.

2.3.3 *Flow Tests with Different Particle Sizes*

In the next part of the feasibility study, glass particles of different sizes in the range of 20–40 μm and of the size of 106 μm were tested. They were suspended in water by mixing using two types of mixers: the magnetic stirrer rotating with the speed of 0.33 m/s (for the 106 μm sized particles), and the high-shear mixer rotating with the speed of 1.56 m/s (for the 20–40 μm particles). The results corresponding to the particle mass fraction of $2\%_{wt}$ are plotted in Fig. 2.24. A reference signal was acquired for pure water flow, in order to calculate the relative change for signals acquired for the particles. This means that the amplitude acquired for water represents $100\%_{rel}$ and the amplitude acquired for the particles is smaller than $100\%_{rel}$. The frequencies below 10 MHz are within the useful sound absorption region.

A detailed analysis of sound spectra is performed by Wrobel and Time (2012), where various effects influencing the spectra, including the particle size, are discussed. The acoustic theories that relate to the critical frequency and the particle interactions, according to the viscous and thermal loss theories, which are dependent on the particle size are discussed in Sect. 2.5.3.

The particle size parameter influences the critical frequency, at which the maximum sound loss occurs. The critical frequency is approximately 8.9 MHz for the 20–40 μm particles, and 6.3 MHz for the 106 μm particles. This is because the small particles attenuate mostly high sound frequencies, above 10 MHz (lower frequencies passed through the particles without the significant energy loss). This effect relates directly to the particle size. The big particles attenuate mostly low sound

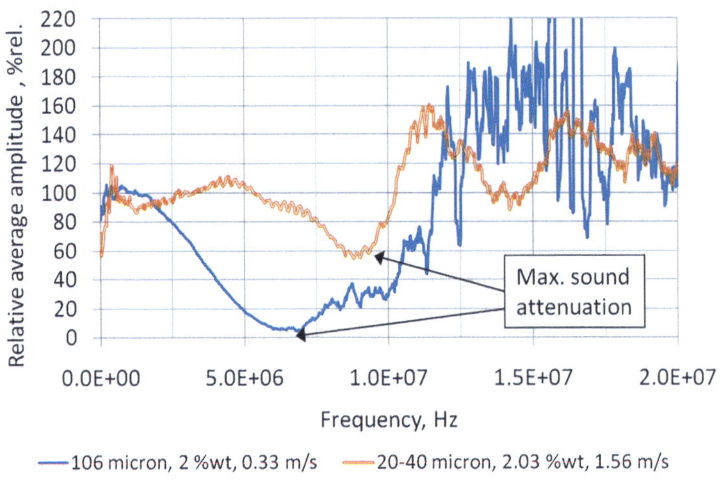

Fig. 2.24 The sound attenuation spectra of the flowing water-glass particle suspension (20–40, 106 μm), for the mixer speeds of 1.56 and 0.33 m/s.

frequencies, lower than 10 MHz. This is confirmed by the results obtained by Dukhin and Goetz (1996). The particle size is related to the critical frequency, however, only for the case of no particles interactions (for a dilute system). For more concentrated systems, the particle volume fraction influences the critical frequency, and it cannot be related directly to the particle size, but to the interparticle distance, which is smaller than the particle size. This is the case in the tests, where the calculated particle sizes (Eq. 2.14, Sect. 2.5.3) are 0.12 μm and 0.14 μm for the 20–40 μm and the 106 μm particles, respectively. The interparticle distance is not only influenced by the high particle fraction, but also by the flow itself.

2.3.4 Conclusions

- The improved pulsed broadband ultrasonic attenuation spectroscopy is suitable for testing of liquids having different physical properties, through the study of time signals and their frequency spectra.
- The test results showed that the liquids of high viscosity attenuate the ultrasonic signal to a higher degree than of low viscosity. This agrees with many examples available in the colloid literature.
- The output ultrasonic signal contains the same number of frequencies as the input signal. This is in contradiction with the simulation results (Sect. 2.2.9). However, the medium response effect is clearly distinguishable through larger attenuation of the signal's frequency peaks, Figs. 2.22 and 2.23.
- The spectroscopy method is sensitive to particle sizes, Fig. 2.24.
- To measure particle sizes in liquid-particle flow using the spectroscopy method, advanced signal interpretation is needed.

2.4 The Ultrasonic Measurement of Liquid-Solid Particle Flowing Suspensions

2.4.1 Subproject Strategy, Objectives, Challenges, Procedure

2.4.1.1 The Strategy for the Ultrasonic Testing of Particle Suspensions

Achieving online acoustic spectroscopy measurement of flowing particle suspensions is in an ongoing development phase. Available published results indicate on utilizing different ultrasonic techniques for that purpose. For instance, listening to the acoustic signals produced by fluid flow with particles, Haugen and Hodgson (1995), the analysis of multiple reflections and sound impedance at a single sound wave frequency (Bamberger and Greenwood 2004). Ultrasonic techniques have been traditionally used for offline measurements of concentrated slurries and emulsions (International Organization for Standardization 2006). Example applications are cement slurries, ceramics, food products and cosmetic emulsions (Dukhin and Goetz 2002).

Measurement of liquid-particle flow in process applications poses challenges, especially when it comes to online measurement of particle concentration. Measurement of dynamic processes, such as flow of liquid transporting particles is indeed challenging, due to its properties changing rapidly in time. This dynamics can be measured by means of the concentration profile representing the local distribution of particles, and the size distribution profile representing the local distribution of various size particles. The ultrasonic attenuation spectroscopy technique has potential for such characterization. The measurement will be affected by the flow instabilities, created by flow dynamics and flow turbulence structures. Therefore, broadband and at the same time fast measurement is essential in such cases. This condition enforces an incident ultrasonic wave to be as short as possible, enabling a short measurement time.

The motivation for this project is to elaborate the methodology that has the potential for online measurement of liquid-particle flow systems, using the ultrasonic attenuation spectroscopy. Various analyses have been performed, forming the methodology: (1) The influence of traditional and novel broadband pulse shapes on the quality of their frequency spectra. (2) Testing the instrumentation's performance in terms of the limitations and measurement sensitivity. (3) Studying the phenomenon of the signal resonance in the square-shaped cell and its influence on measurements. (4) The numerical study of the square pulse design (in MATLAB), including the quality improvement of the frequency spectra, by regulating the lengths of the rise/fall times and the number of points in its vertical and horizontal sections. (5) Techniques for signal acquisition (accounting for the pulse repetition frequency, the number of wave cycles) and signal averaging (the ensemble and running averages).

In the next steps, the following studies will be performed: (6) The measurement of the particle concentration in a low concentration range for flowing water-glass beads suspensions. (7) Analyzing the equation of motion for a damped system under harmonic force. (8) Introduction into the acoustic theories for particulates. (9) Testing of concentrated particle suspensions in high-shear flow.

2.4.1.2 The Subproject Objectives

- The measurement of the particle concentrations in a low concentration range at flow conditions using the ultrasonic attenuation spectroscopy technique.
- Finding a suitable theory and reference examples. A thorough analysis is not intended at this stage, however, it is performed in Sect. 2.5.
- Finding the sensitivity of the method to small concentration changes.

2.4.1.3 The Subproject Challenges

- Choosing proper parameters of the incident ultrasonic wave (the shape, the amplitude, the pulse repetition frequency, the number of cycles), which are compatible and consistent with used instruments in the test (the signal generator, the data acquisition card, the 5 MHz ultrasonic transducer), to ensure the high-accuracy measurement of liquid-particle flow.
- Choosing suitable particles for tests. The best particles have high-density contrast compared to liquid, in order to ensure the acoustic transparency.
- Establishing small enough concentration steps (measurements are possible only up to the maximum concentration, due to signal damping).
- Ensuring a homogenous distribution of glass beads.
- The acquisition of signals according to the established principles (Sect. 2.2.4).
- Performing signal averaging in the post-processing phase.
- Studying correlations between acquired ultrasonic signals and particle concentrations.

2.4.1.4 The Procedure

A test was performed for the flow of water with the glass beads (ϕ = 106 µm, ρ = 1.56 g/cm^3), for the mixer speed of 16 RPM (0.33 m/s). The magnetic stirrer was used for the flow generation in the square test cell (Fig. 2.2, Sect. 2.2.2). The particle mass fraction varied from 0_{wt} to $1.25\%_{wt}$, every $0.25\%_{wt}$. The square pulses (10 kHz PRF, 5 V) were used as the input pulses for the ultrasonic attenuation spectroscopy, to obtain the broadband frequency spectra and possibly the shortest pulse (0.1 ms). The test was run in the through-transmission mode. A procedure in detail and explanations are elaborated by Wrobel and Time (2011).

2.4.2 Briefly on the Acoustic Theories for Particulates

Several acoustic theories for particulates have been developed, taking into consideration the sound absorption and the sound scattering mechanisms (Sect. 2.5.3). They are all complementary and still in an ongoing development phase, describing various aspects and applications. They have been developed, based on studies of different types of concentrated particle suspensions, and majority of them function under the assumption of dilute suspensions (no particle interactions present). Practical ultrasound technology applications are challenged with the particle sizes ranging from nm to mm, dilute and high fractions (promoting the particle interactions), and complex liquids flowing at any speed. Available for measurement ultrasound frequencies range from 20 kHz up to typically 40 MHz, possibly up to several gigahertz. Empirical solutions come into play in practical applications. The physical theory of the damped, driven harmonic oscillator describes the interactions between a wave and a liquid-particle system, and is complementary to the theories presented in Sect. 2.5.3.

2.4.3 The Damped Driven Harmonic Oscillator Theory

The generation and propagation of ultrasounds in suspensions can be described by the physical theory of the damped, driven harmonic oscillator. Harmonic oscillators related to studied applications are the active element of an ultrasonic transducer and the tested suspension. The equation of motion of the damped, driven oscillator is based on Newton's second law and is given by Rao (1995):

$$m\frac{d^2x}{dt^2}+c_v\frac{dx}{dt}+kx=F_0\cos(\omega t) \tag{2.5}$$

where:

F_0—the driving amplitude of the harmonic force (V)
ω—the driving frequency of the incident ultrasonic wave, $\omega = 2\pi f$
m—the mass
c_v—the viscous damping coefficient
k—the system stiffness coefficient (the inverse of the compressibility)

The steady-state solution of the equation of motion, $x(t)$, corresponding to acquired time signals in the spectroscopy tests, is also harmonic, Rao (1995):

$$x(t)=X\cos(\omega t-\phi) \tag{2.6}$$

where:

X—the amplitude

ϕ—the phase angle

The frequency of the oscillator is the same as that of the driving force. There will be no extra frequencies in an acquired signal, after it passes through a sample (Sect. 2.2.9). The amplitude and the phase angle parameters are scaled by the amount that depends on the driving force frequency, in relation to the resonant frequency of an oscillating system and the damping ratio, Rao (1995):

$$\frac{X}{\delta_{\mathrm{st}}} = \frac{1}{\sqrt{\left(1-\left(\frac{\omega}{\omega_{\mathrm{n}}}\right)^2\right)^2 + \left(2\xi\frac{\omega}{\omega_{\mathrm{n}}}\right)^2}} \tag{2.7}$$

$$\phi = \tan^{-1}\left(\frac{2\xi\frac{\omega}{\omega_{\mathrm{n}}}}{1-\left(\frac{\omega}{\omega_{\mathrm{n}}}\right)^2}\right) \tag{2.8}$$

where:

ω_{n}—the resonant frequency of an oscillating system, $\omega_{\mathrm{n}} = (k/m)^{1/2}$
ξ—the damping ratio, $\xi = c_{\mathrm{v}}/(2m\omega_{\mathrm{n}})$
δ_{st}—the deflection under static force, $\delta_{\mathrm{st}} = F_0/k$

A theoretical analysis was performed to illustrate the characteristics of the mechanism of a wave interaction with a suspension, which is qualitatively represented by the amplitude and the phase shift of the signal, Fig. 2.25. Behaviors of the signal amplitude and the phase offset were analyzed for three cases:

- A decrease of c_{v} for constant values of m and k.
- An increase of m for constant values of c_{v} and k.
- A decrease of k for constant values of m and c_{v}.

Additionally, it was assumed that the viscous damping coefficient, c_{v}, and the system stiffness coefficient, k, are not dependent on the mass, m. In real situation, the viscosity of a suspension varies exponentially with particle fraction (Hunter 1993). The stiffness of a suspension increases (and the compressibility decreases) with increasing particle fraction (Urick 1947).

The study has showed that when the viscous damping coefficient, c_{v}, was reduced, the signal amplitude, X, increased and the phase shift, ϕ, diminished. Whereas the increased mass, m, caused either the increased or decreased amplitude, X, and the phase shift, ϕ (the amplitude and phase shift points moved to the left), depending on whether the resonant frequency of a system, ω_{n}, was smaller or larger than the driving force frequency, ω. The same phenomena occurred for the decreasing stiffness coefficient, k.

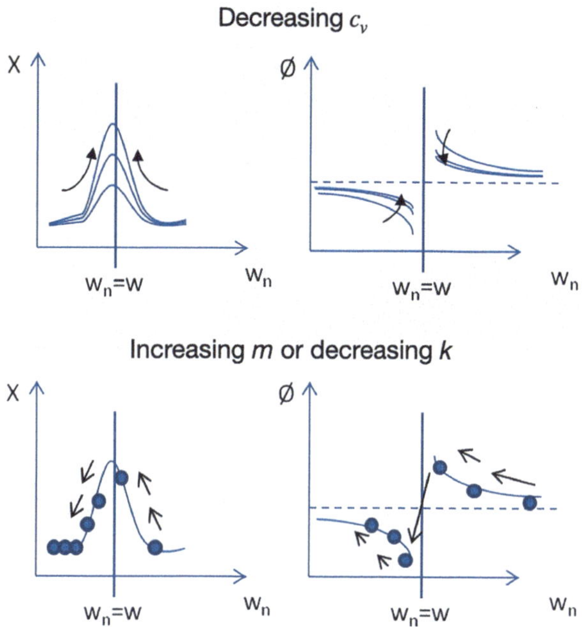

Fig. 2.25 The analysis of the amplitude and the phase shift, which depend on the system resonant frequency, according to the harmonic oscillation equation. (B.M. Wrobel, R.W. Time, *Applied Acoustics* 72 (2011) 324–335).

2.4.4 Particle Concentration Measurement Through the Sound Attenuation

Following the test procedure in Sect. 2.4.1, measurements were performed. The acquired time signals for water-glass beads flow are shown in Fig. 2.26a. Three major effects can be observed for the increasing particle fraction up to $1.25\%_{wt}$:

1. The signals' phase shifts in time domain, which are used for the sound speed measurements. The phase shift effect is proportional to the mass of particles and is dependent on the viscosity and compressibility of the suspensions, according to Eq. (2.8).
2. The amplitude attenuation caused by the increasing particle fraction.
3. The change of the signals' shapes. It is the effect caused by diverse attenuation rates of different wave harmonics, caused by the varying particle mass fractions.

The frequency spectra in Fig. 2.26 show the sound attenuation dependent on the average particle mass fraction. The attenuation increases for the increasing fraction (Figs. 2.26 and 2.27). The waviness of the spectrum in Fig. 2.26b is due to the resonance effect (the wave interference) in the wall of the test cell.

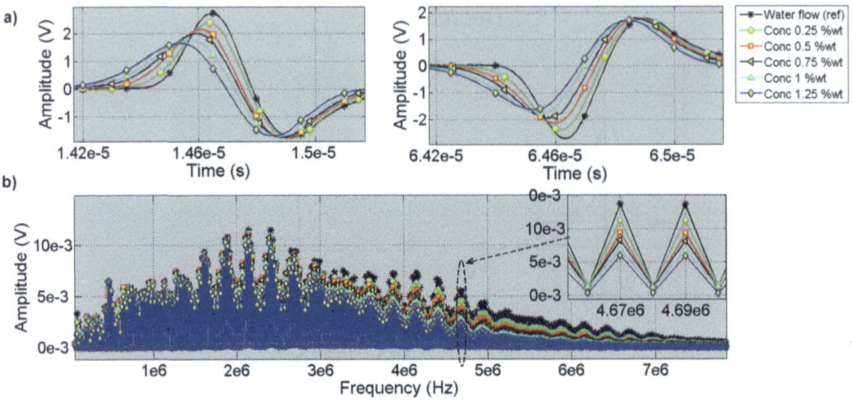

Fig. 2.26 (**a**) The instantaneous pulsed broadband ultrasonic attenuation spectroscopy signals (the square pulses) for water and water-glass particle flow, for the particle mass fractions of 0.25–1.25%$_{wt}$ and the mixer speed of 0.33 m/s (the magnified left and right parts of the square pulses). (**b**) The corresponding frequency spectra. (B.M. Wrobel, R.W. Time, *Applied Acoustics* 72 (2011) 324–335).

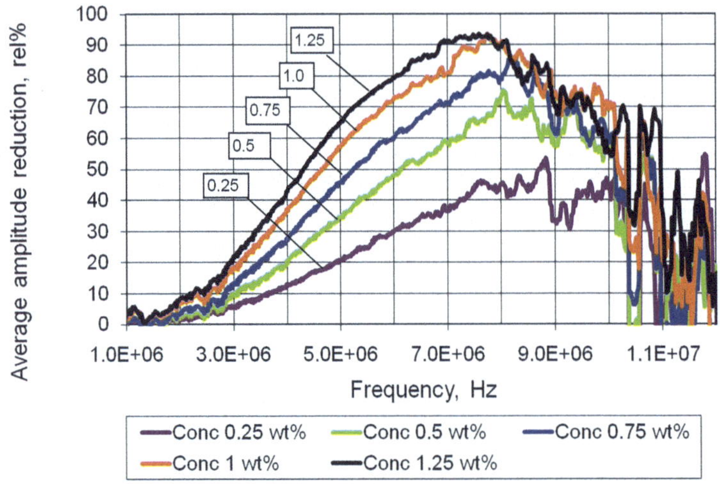

Fig. 2.27 The relative amplitudes of the ultrasonic square pulses for water-glass particle flow, for the particle mass fractions of 0.25–1.25%$_{wt}$, and the mixer speed of 0.33 m/s. (B.M. Wrobel, R.W. Time, *Applied Acoustics* 72 (2011) 324–335).

2.4.5 Conclusions

- The experimental test for the flows of water and water with glass particles, for the mass fractions of 0.25–1.25%$_{wt}$, varying by the step of 0.25%$_{wt}$, demonstrated that the improved pulsed broadband ultrasonic attenuation spectroscopy technique is a sensitive and an accurate tool, with the potential for online measurement of liquids and liquid-particle suspensions.

- The wave interference can be observed in the spectra. It can be minimized by the avoidance of the ultrasonic wave reflecting back and forth between the parallel walls of the test cell. However, this kind of noise is not harmful, unless it significantly dampens the amplitude.
- The equation of the damped system under harmonic force is suitable for describing the ultrasonic measurement of liquid-particle flow. The properties of a suspension (the mass, viscosity, and stiffness) can be translated into a wave, for the known properties of the incident wave (the amplitude and frequency). However, the equation does not include all important parameters that need to be considered in a complete analysis. These parameters are, for example, the inertial effect resulting from different motions, as well as the velocities of liquid and particles, and the temperature change.
- It is advised to produce high measurement statistics and calibrate signals, due to the signal noise and system instability effects, which are inherent with the dynamic nature of flow, affecting online measurements.

2.5 The Ultrasonic Measurement of Liquid-Particle Flow in High Shear

2.5.1 Subproject Strategy, Objectives, Challenges, Procedure

Online process metering involving multiphase flow measured by ultrasound instrumentation is challenging, not only due to liquid, gas and solid phases flowing simultaneously in a pipeline, but also due to flow turbulence. According to Endress+Hauser (2006) turbulent flow exists in 95% of closed pipelines; therefore, flowmeters should normally be designed to operate under such flow conditions.

The turbulence effects that are created in high-shear flow have significant influence on the attenuation and dispersion of sound. Even in pure liquid without any particles, the flow turbulence intensity will play an important role in the sound propagation through it: “only recently, some of ultrasonic Doppler flowmeters have been developed to work with clean liquids, reflecting off turbulent swirls” (Winkler 2010). The characteristic features of a liquid-particle mixture flow in high shear are the multiple particle-particle interactions, Harker and Temple (1988) and the presence of turbulent eddies (Rabenjafimanantsoa 2007). These factors may make a difference when particle fractions are measured using ultrasound and they may cause measurement errors, if they are not accounted for. There is a large unexploited area, when it comes to the ultrasonic measurement of flowing suspensions in turbulent flow conditions.

The motivation for this study is to verify the suitability of the pulsed broadband ultrasonic attenuation spectroscopy for the simultaneous measurement of the average particle concentration and average flow speed parameters, using a single ultrasonic pulse. The spectroscopy method has been complemented with the sound speed measurement. Another motivation is to check the potential for online flow measurement, in high-shear and turbulent conditions, using the elaborated methodology. Finally, it is desired to investigate possibilities for the characterization of liquid-particle flow, the control of other flow dynamics parameters. It is important to underline that available research work with respect to the goal of achieving online acoustic spectroscopy measurement of flowing particle suspensions is under ongoing development.

The following studies will be performed: (1) The studies of the acoustic theories for particulates and sediment transport theory, together with relevant published results. (2) Sound attenuation and sound speed tests for flowing water with glass beads in high shear. (3) The conversion of measured time signals into the frequency domain and the analysis of the frequency spectra. (4) Analyzing the standard deviation plots in terms of the flow process stability. (5) Temperature tests for the sound attenuation and the sound speed measurements in water and the acrylic wall of the test cell. (6) The evaluation of the measurement accuracy. (7) Analyzing the measurement uncertainty. (8) The study of particle clustering and air bubbles in the flow of water with glass beads, by ultrasonic measurement.

2.5.1.1 The Hypothesis for the Sound Loss Due to the Flow Turbulence Intensity

A hypothesis to be tested assumes that the flow turbulence intensity, $\varepsilon(Re)$, may significantly influence any ultrasonic measurement of concentrated particles in liquid. An associated theory describing the interaction of an ultrasound pulse with colloids is the superposition theory by Dukhin and Goetz (1996, 2002, 2017). According to this theory, the total attenuation of sound (the sound loss) is the result of the contribution of several mechanisms: the viscous, the thermal, the structural, the scattering, the intrinsic, and the electrokinetic mechanisms. The electrokinetic mechanism is of negligible contribution to the total sound loss (Dukhin and Goetz 1996). An additional sound loss, namely the turbulent flow sound loss α_{turb}, is expected due to the flow turbulence intensity in turbulent flow applications.

2.5.1.2 Comments on the Acoustic Models for Particulates

Early acoustic models for liquid-particle suspensions, namely the phenomenological models by Urick (1947) and Ament (1953), the coupled phase models by Harker and Temple (1988), Gibson and Toksöz (1989), have not been designed to account directly for the turbulent effects in a liquid-particle mixture, in terms of the presence of the inertia forces of liquid and particles in high-shear flow. They are related to different velocity values of the liquid and the particles and their high Reynolds numbers (described by non-Stokes' law). There are also other turbulent effects, such as the effects associated with the nonlinear distribution of particles in liquid, creating vortexes in flow that trap particles, according to the sediment transport theory, Nielsen (1992), as well as the self-organized motion of particles connected in structures (Nielsen 1992; Rabenjafimanantsoa 2007). In the acoustic models for liquid-particle suspensions, the simple Stokes flow is often utilized, however being only valid for laminar flow with the Reynolds numbers smaller than one. Stokes' drag for steady motion is also introduced in the models and the distribution of solid spheres is assumed to be statistically homogenous.

Such approximations are useful, when the propagation of a sound wave through a steady flowing system (without changing properties in time) is analyzed, however, for a suspension consisting of heavy particles flowing at high speeds, these assumptions do not apply. It also seems that the closure relations for turbulent flow, Manninen and Taivassalo (1996), are missing in the acoustic models for particulates and therefore they cannot be applied for practical online liquid-particle flow measurement applications. Another limitation of the acoustic models for sound attenuation is validity only for a dilute suspension, without consideration of particle interactions in it (Temkin 1998).

2.5.1.3 The Subproject Objectives

- To perform tests with low concentration liquid-particle suspensions in high-shear (turbulent) flow, using the pulsed broadband ultrasonic attenuation spectroscopy method.
- To verify the method's suitability for the simultaneous measurement of the average particle concentration and average flow speed.
- To verify the method's suitability with respect to other flow dynamics parameters.

2.5.1.4 The Subproject Challenges

- The assurance of stable flow and measurement conditions (a suitable position of the mixer, good adhesion of the transducers).
- Performing the measurements of flow of suspensions for turbulent flow velocities, using two independent ultrasonic techniques, one utilizing the sound attenuation measurement and the other utilizing the sound speed measurement, at controlled temperature conditions, according to the established measurement principles (Sect. 2.2.4).
- Observing air bubbles present in the system, flowing at high mixer speeds.
- Observing particle behavior during the flow tests.
- The post-processing of acquired big data signals (using the developed programs in MATLAB).

- Studying signal behavior for changing particle concentration and flow speed parameters.

2.5.1.5 The Procedure

An ultrasonic test of a concentrated liquid-particle system in high-shear flow conditions was carried out using the spectroscopy technique. The glass particles (ϕ = 20–40 µm, $\rho = 2.47$ g/cm^3) in water, flowing at the mixer speeds of 1.5–2 m/s, were studied for the mass fractions up to 10%$_{wt}$ (5%$_{vol}$). The high-shear mixer and the square test cell were used in the test (Fig. 2.3, Sect. 2.2.2). The test was performed for a uniform distribution of particles in water, therefore, the homogeneity was assumed. The square pulses (10 kHz PRF, 5 V) were used for the ultrasonic spectroscopy and the sound speed tests, while the spike pulses (500 Hz, 100 µJ) were used only for the sound speed test. Particle interactions were expected in the test, due to turbulent flow conditions. Additional temperature tests for the sound attenuation and the sound speed measurements of water and the acrylic wall were performed, for the temperature range of 20–40 °C and supported by numerical simulations. The procedure and explanations in detail are elaborated by Wrobel and Time (2012).

2.5.2 Experimental and Interpreting Results Designs

In order to obtain reliable experimental results, several studies were performed, Fig. 2.28:

1. The hardware and software elements of the experimental setup were prepared and tested. The instrumentation was tested in terms of the measurement sensitivity and its limitations, in order obtain a high-quality signal for ultrasonic measurement of liquid-particle flow.
2. Signal acquisition methods were studied, to obtain a high-accuracy signal. Successful signal acquisition is fulfilled by the high-resolution waves containing a high number of points per cycle, acquired full wave cycles, and by good data statistics ensured by the measurement repeatability, which is important for flow data constantly changing in time. A suitable pulse repetition frequency should be chosen to fit with the ultrasonic transducer's frequency. Suitable wave frequencies should be chosen for the detection of small particles, with the wavelength that captures the particles well.
3. Various pulse types (the sine, triangle, square, spike, sweep and burst waves) and designs (with respect to the density of points, frequency and symmetry) were studied. It turned out that the square wave type was particularly interesting, due its harmonic nature, with its multiple frequencies available for particle measurement.

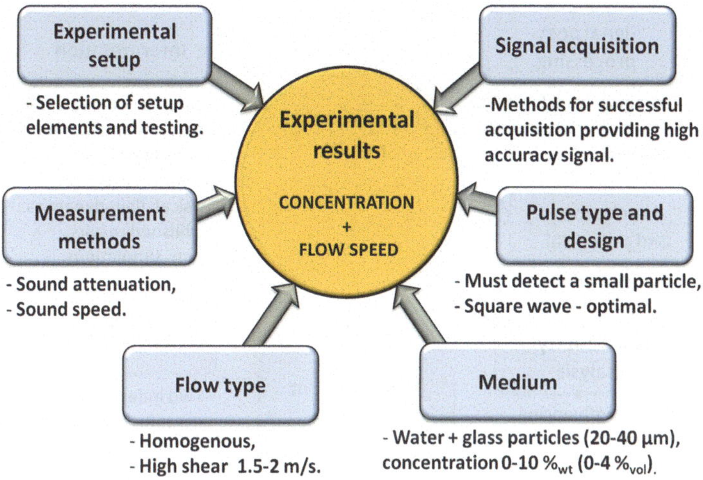

Fig. 2.28 The design for the ultrasonic experiments.

4. The glass beads of the sizes of 20–40 μm were chosen for testing. Larger size particles were too heavy and easily sedimenting on the cell's bottom during the test, which was not desired. The small particles were well-suspended in the flow and created a homogenous mixture, therefore, represented true average particle concentration in the cell. The particle concentration was chosen to be low, because it was not possible to measure high concentrations due to the substantial sound attenuation.
5. The homogenous type of flow regime was desired. This was obtained by running the mixer at 1.5–2 m/s.
6. Two measurement methods were used: the sound attenuation using the spectroscopy as the main method, and the sound speed as the supportive method.

After experimental data were collected, the results were analyzed, Fig. 2.29:

1. Datasets were averaged and converted from the time into the frequency domain using the fast Fourier transform. Operations on the big datasets were computing memory-demanding and required advanced programming codes. These programs consisted of loops, subloops, memory-clearing commands, and multivariable matrixes (arrays).
2. After completing the signal processing and plotting, the signal was interpreted. As several effects were measured at once using only one signal, therefore the straightforward interpretation was challenging. The signal was influenced by the increasing particle concentration and flow speed, the flow vortexes, the changing system temperature, the sporadical gas bubbles and the particle clustering, which occurred during the experiments. In order to separate individual effects, the acoustic, the physical, and the flow dynamics theories were studied, together with published results available in the literature. Numerical simulations were performed supportively.

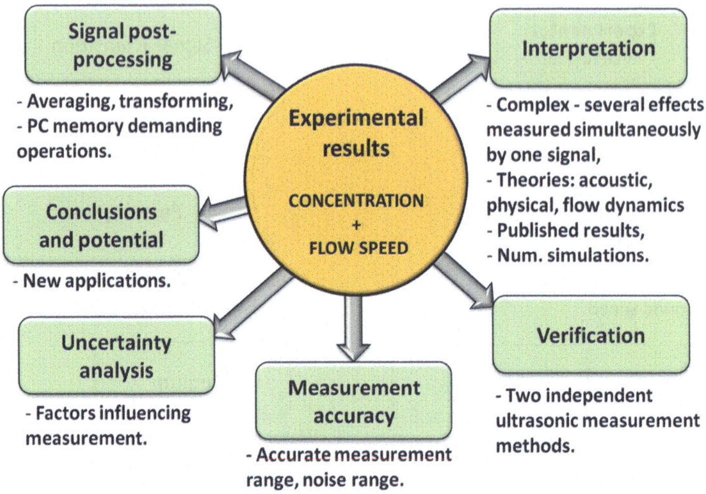

Fig. 2.29 The design for the results interpretation.

3. The verification of the experimental results was performed by two independent ultrasonic methods, and by comparing them with published results.
4. The analysis of the signal resulted in finding the measurement and noise frequency ranges.
5. The uncertainty analysis was performed, stating factors influencing the measurements. For example, the particle concentration, the fluid speed, the presence of air bubbles, the system temperature.
6. Final conclusions and potential for new applications were stated.

2.5.3 *The Acoustic Theories for Particulates*

2.5.3.1 The Sound Attenuation Through the Absorption and Scattering

Ultrasound attenuation in colloids arises either from the sound absorption (the acoustic energy conversion into the thermal energy) or the sound scattering (the acoustic energy redirection).

$$\text{Sound Attenuation} = \text{Absorption} + \text{Scattering}$$

For ultrasounds, the absorption and scattering regions are distinctly separated and visible in the frequency spectra. As a rule, for a given particle size, the absorption of ultrasounds is dominant at low frequencies, whereas the scattering is dominant at high frequencies, with the overlapping region existing within the narrow intermediate frequency range.

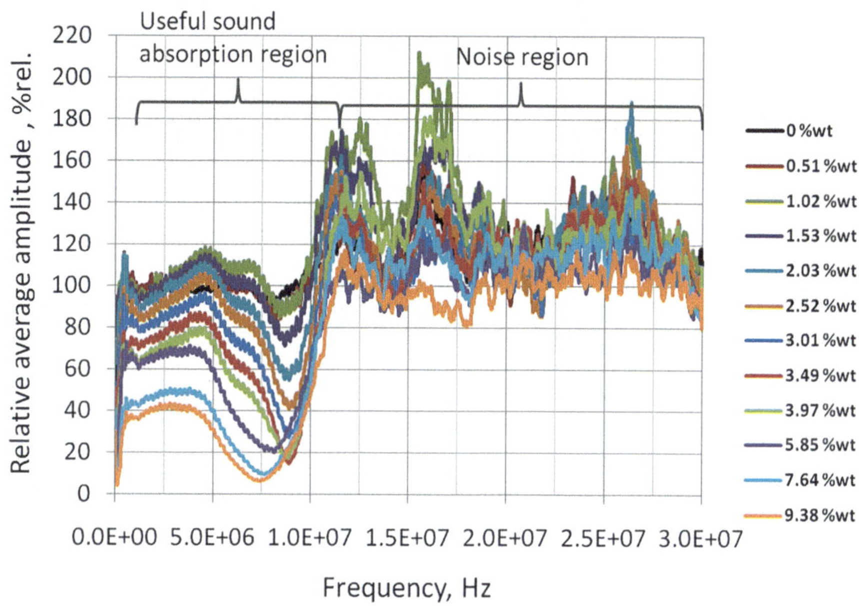

Fig. 2.30 The relative sound attenuation for the particle mass fractions 0–9.38$\%_{wt}$ in water, for the mixer speed of 1.56 m/s. (B.M. Wrobel, R.W. Time, *Applied Acoustics* 73 (2012) 117–131).

In the tests, the concentrated suspensions were measured only within the sound absorption region up to 12 MHz, Fig. 2.30. The noise begins around 12 MHz and is present at higher frequencies. The useful frequency range is within 2–12 MHz and the critical frequency is at the maximum sound attenuation (7.5–9 MHz). The natural frequency of the 5 MHz transducer imposes a limitation to detect the sound of small amplitudes at higher frequencies (Fig. 2.11, Sect. 2.2.5).

2.5.3.2 The Critical Frequency of the Maximum Sound Attenuation

A frequency at which a sound wave reaches the maximum attenuation is called the critical frequency, ω_{cr} (Dukhin and Goetz 2002; Gibson and Toksöz 1989). It is related to an instance of the particle size or the interparticle distance, from which the latter one depends on the particle volume fraction.

2.5.3.3 The Phenomenological Model by Harker and Temple (1988)

The phenomenological model by Harker and Temple (1988) is the simplest approach for calculating the velocity and attenuation of ultrasounds in suspensions. The sound speed of the suspension of fluid with particles, c (m/s), can be calculated by the following equation:

$$c = \frac{1}{\sqrt{\rho_{\text{eff}} \kappa_{\text{eff}}}} \tag{2.9}$$

where ρ_{eff} (kg/m^3) is the effective mixture density and κ_{eff} (Pa^{-1}) is the effective mixture compressibility. According to Urick (1947), these parameters are calculated based on the linear averaging of two media properties using the volume fraction of dispersed particles, φ (dimensionless):

$$\rho_{\text{eff}} = \varphi \rho_{\text{p}} + (1 - \varphi) \rho_{\text{m}} \tag{2.10}$$

$$\kappa_{\text{eff}} = \varphi \kappa_{\text{p}} + (1 - \varphi) \kappa_{\text{m}} \tag{2.11}$$

The index m refers to the dispersion medium (liquid) phase and the index p refers to the dispersed phase (particles). There are several other equations for ρ_{eff} and κ_{eff} that are dependent on a larger number of parameters, than only the volume fraction of dispersed particles. For example, Ament (1953) included the effects of fluid viscosity and particle size. Biot (1956), cited by Harker and Temple (1988), considered the effect of tortuosity (the tortuosity is commonly used to describe the diffusion in porous media); however, this theory is only valid for the suspension having large volume fraction, which becomes more akin to a porous solid.

The sound attenuation is expressed by the particle fraction, φ, multiplied by the "average" scattering factor, σ, Gubernatis and Domany (1984), cited by Harker and Temple (1988):

$$\alpha = \varphi \sigma \tag{2.12}$$

2.5.3.4 The Coupled Phase Model for Dilute Systems by Gibson and Toksöz (1989)

The coupled phase models provide the best framework to describe the ultrasound absorption. They are solved through the equations of motion and the equations of continuity, each for fluid and solid particles and the equation for the drag on one phase by the other. Gibson and Toksöz (1989) presented the coupled phase model valid for the concentrations not greater than 0.20. The authors proposed two different equations for the attenuation coefficient, valid for low and high sound frequencies. The model is based on establishing the critical frequency, ω_{cr}, corresponding to the maximum attenuation that separates the frequency ranges. The inertial effects are negligible in the low frequency range, however, they play a significant role in the high frequency range.

2.5.3.5 Temkin's Model (1998)

Temkin (1998) considered fluid and particles moving with different velocities and the heat transfer between them. The heat transfer results in the irreversible energy transfer, affecting the suspension's compressibility. Based on these premises, the model was built for the determination of attenuation and speed sound in a flowing particle suspension. The obtained equations define attenuation and speed sound for a colloidal system in terms of the particle-to-fluid velocity and the temperature ratios, within a wide sound frequency range. The model is limited to dilute suspensions (no interactions between particles are assumed).

2.5.3.6 The Multiple Scattering Theory by McClements (1991)

McClements (1991) described the ultrasonic propagation in scattering systems, where the sound speed and attenuation are dependent on the particle size and concentration, through the visco-inertial and thermal properties of particles and a liquid medium. The single scattering theory is applicable only to dilute systems, while the multiple scattering theory takes into account the total scattering from all particles, with their spacial distribution in a system, where the significant amount of energy scattered from one particle is incident upon neighboring particles (McClements 1991). This portion of the sound does not reach the receiving transducer; therefore, the scattering mechanism attenuates the sound, in addition to the sound absorption. The effective density and effective compressibility of liquid-particle systems are important for multiple scattering systems. The overall attenuation due to scattering increases with the sound frequency, Fig. 2.30.

2.5.3.7 The Superposition Theory by Dukhin and Goetz (1996, 2002, 2017)

The superposition theory proposed by Dukhin and Goetz (1996) is a combination of the scattering theory and the coupled phase model, where six different mechanisms of sound interaction with colloids are distinguished: the viscous, the thermal, the structural, the scattering, the intrinsic, and the electrokinetic mechanisms. The contribution of the electrokinetic mechanism to the total acoustic attenuation is considered to be negligible (Dukhin and Goetz 1996). The total sound attenuation, α_T (Np/m), is expressed as (Dukhin and Goetz 2017):

$$\alpha_T = \alpha_{vis} + \alpha_{th} + \alpha_{st} + \alpha_{sc} + \alpha_{int} \tag{2.13}$$

The viscous losses are related to the shear waves generated by particles oscillating in liquid, in the acoustic pressure field (the particles move with different motion than the liquid, due to the difference in their viscoinertial properties); the shear waves appear due to a high density difference between the particles and the liquid, McClements (1991) and are important for $\Delta\rho/\rho > 1$ (Dukhin and Goetz 1996). In the conducted test with the glass particles (2.47 g/cm^3) and water (1 g/cm^3), the density contrast is equal to 1.47, therefore, the viscous losses are important to account for. *The thermal losses* are related to the temperature gradients generated near the particle surface, McClements (1991), and they are dominant for soft emulsion and latex droplets (Dukhin and Goetz 1996). The thermal losses are important up to $ka = 0.5$, where k (1/m) is the wavenumber and a (m) is the particle diameter (Anson and Chivers 1990, cited by Dukhin and Goetz 2017). In the case of *the scattering losses,* the particles redirect a part of the acoustic energy, resulting in the amount of sound that does not reach the receiving transducer (Fig. 2.52 in Sect. 2.6.2, Fig. 2.55 in Sect. 2.6.4). The scattering losses are dominating for the large particles with $ka > 1$, Dukhin and Goetz (2017), for example, for the ultrasound frequencies above 15 MHz, Stakutis et al. (1995). *The intrinsic losses* are the losses of acoustic energy due to the interaction of sound waves with the materials of the particles and the medium individually, on their molecular level (Stakutis et al. 1995; Dukhin and Goetz 1996; Ainslie and McColm 1998). Stakutis et al. (1995) considered the viscous, the thermal, the intrinsic and the scattering losses in their work with the ultrasound attenuation in aqueous suspensions. *The structural losses* occur, when the particles are joined together in some network. The oscillation of these interparticle bonds can cause additional energy dissipation (Dukhin and Goetz 1996, and Fig. 2.43 in Sect. 2.5.8). *The electrokinetic mechanism losses* are related to the oscillations of the charged particles in the acoustic field (Richards 1931; Dukhin and Goetz 1996; Davidson et al. 2007). All of the mechanisms, excluding the electrokinetic mechanism, will contribute to the sound attenuation in the studied cases of glass beads-water suspensions. There is one more element extending the theory, namely *the flow turbulence losses*, α_{turb}, which are the sound losses occurring due to turbulent flow structures (Fig. 2.37 in Sect. 2.5.7, Fig. 2.44 in Sect. 2.5.8, and the seventh general conclusion in Sect. 2.5.10).

2.5.3.8 Particle Interactions According to the Viscous and Thermal Theories

The viscous loss theory assumes a parameter called the viscous layer thickness, δ_{n} (m). The particle is surrounded by the viscous layer with the thickness δ_{n}. According to the theory created by Epstein and Carhart (1953), Allegra and Hawley (1972)—the ECAH, cited by Dukhin and Goetz (2002, 2017), the work of Harker and Temple (1988) and Mbailassem et al. (2019), it is given by:

$$\delta_{\eta} = \sqrt{\frac{2\eta}{\rho_{\text{m}}\omega}} \tag{2.14}$$

where η is the dynamic viscosity (Pa s), ρ_{m} is the density of a medium (kg/m^3), and ω is the angular sound frequency (rad/s or s^{-1} in the SI base units).

The visualization of the viscous layer thickness, in terms of relations between the critical frequency and particle concentration, for one particle size is presented in Fig. 2.31. When the system frequency, ω, is equal to the critical frequency, ω_{cr}, the

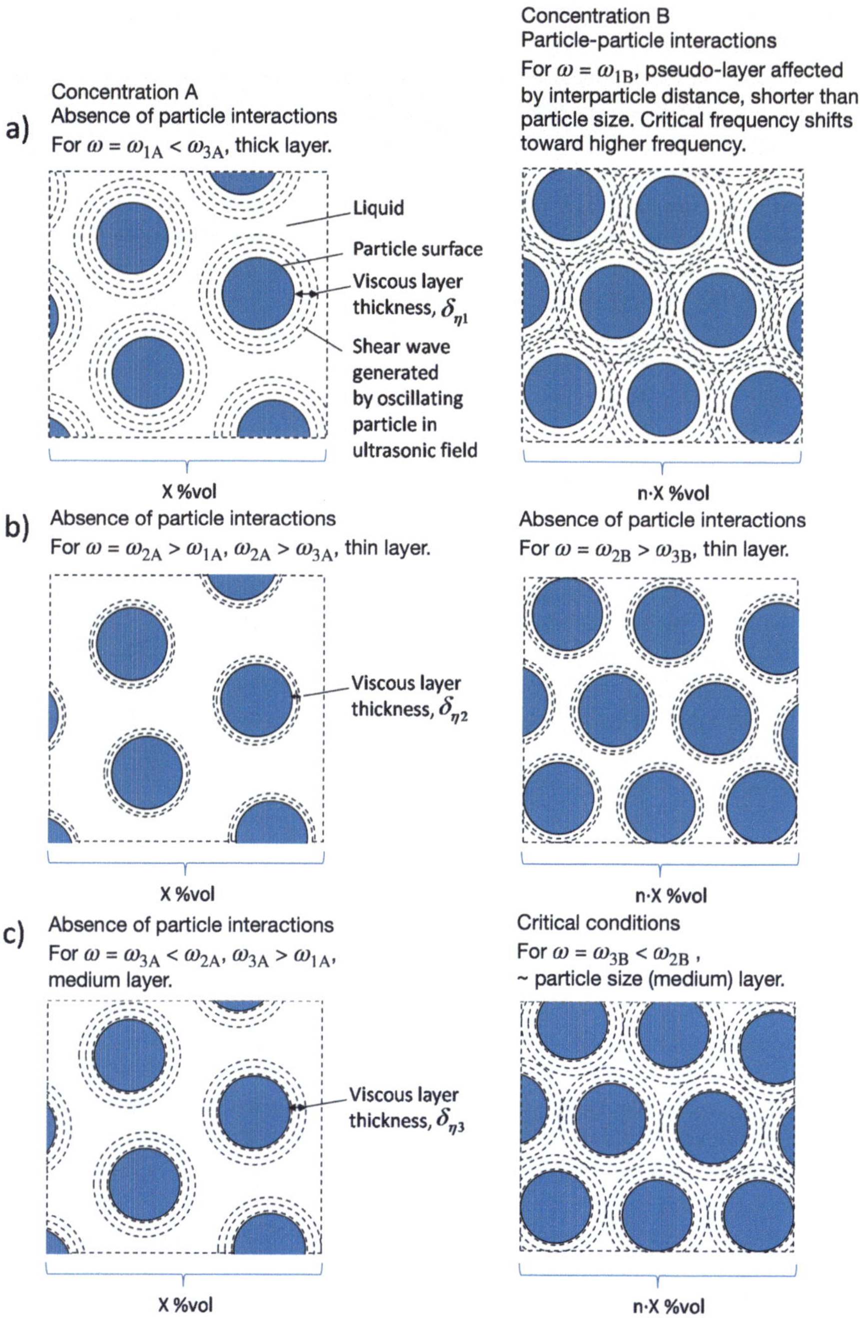

Fig. 2.31 The visualization of the relations between the viscous layer thickness and the system frequency, for particle concentrations (low "A" on the LHS, high "B" on the RHS) and one particle size. (**a**) On the LHS: The absence of particle interactions, the thick layer. On the RHS: Particle-particle interactions, the interparticle distance pseudo-layer. (**b**) On the LHS and RHS: The absence of particle interactions, the thin layer. (**c**) On the LHS: The absence of particle interactions, the medium layer. On the RHS: Critical conditions, the particle size (medium) layer.

particle size can be determined from the relation of $\delta\eta = a$, Dukhin and Goetz (2017), however, this assumption is only valid, when no interactions between particles are present (for dilute suspensions). In the case that particle interactions are present, the interparticle distance parameter overtakes the particle radius as the dominant parameter, and then $\delta\eta < a$, Dukhin and Goetz (2017), which results in the shift of the critical frequency toward the higher sound frequency (Dukhin and Goetz 2017, and the results in Fig. 2.43 for the particle mass fractions up to $3.49\%_{\mathrm{wt}}$, Sect. 2.5.8). The particle interactions are present for non-dilute colloidal suspensions.

Similarly to the generation of shear waves, due to the particles oscillating in a sound wave, the thermal waves are also generated in the system and decay at the distance, δ_{T} (m), due to the sound wave traveling through it. The thermal layer thickness can be calculated from the following equation for colloidal suspensions, Harker and Temple (1988), Dukhin and Goetz (2017), Mbailassem et al. (2019):

$$\delta_{\mathrm{T}} = \sqrt{\frac{2\kappa_{\mathrm{m}}}{\rho_{\mathrm{m}}\omega C_{\mathrm{p}}^{\mathrm{m}}}} \tag{2.15}$$

where κ_{m} is the thermal conductivity of a medium (W/(m K)), ρ_{m} is the density of a medium (kg/m^3), ω is the angular sound frequency (rad/s or s^{-1} in the SI base units), and $C_{\mathrm{p}}^{\mathrm{m}}$ is the specific heat capacity of a medium (J/(kg K)).

The thermal waves could be sketched in a similar way as the shear waves, Fig. 2.31. The effective media density and the compressibility are described by Eqs. (2.10) and (2.11), whereas the equations for the effective thermal conductivity and the heat capacity are presented by Dukhin and Goetz (2017).

2.5.4 Sound Attenuation in Liquids - A Theoretical Study

The total sound attenuation coefficient in liquids consists of the viscous, the thermal, the intrinsic, and the electrokinetic losses. The contribution of the electrokinetic losses to the total acoustic attenuation is considered to be insignificant. There are no scattering and structural losses from the interparticle bonds, which are present for liquid-particle suspensions. Parameters, namely the dynamic (shear) viscosity, Holmes et al. (2011), the bulk (volume) viscosity and the compressibility, Dukhin and Goetz (2009), Zaidni and Benjelloun (2021), the thermal conductivity and the specific heat capacity, Holmes et al. (2011), Zaidni and Benjelloun (2021), the chemical absorption (on the molecular level), Ainslie and McColm (1998), and the electric properties, Richards (1931), Davidson et al. (2007) influence it.

Similarly like for liquid-particle suspensions, there is one more element extending the superposition sound attenuation theory. It is related to the flow turbulence losses, which are the sound losses occurring due to turbulent flow structures (Fig. 2.44 in Sect. 2.5.8, and the seventh general conclusion in Sect. 2.5.10).

The viscous sound attenuation in liquid is given by Stokes' equation (1845), and it is used for the calculation of the viscous sound loss, α_{s} (Np/m), Eq. (2.16). It is

Table 2.8 The calculated viscous and thermal thickness layers in liquids.

Parameter	Water	Glycerol	Kerosene
Viscous thickness layer, δ_n (μm)[a]	5.64	194.59	7.98
Thermal thickness layer, δ_T (μm)[a]	2.15	1.72	1.67
Viscous thickness layer, δ_n (μm)[b]	0.18	6.15	0.25
Thermal thickness layer, δ_T (μm)[b]	0.07	0.05	0.05

[a] At 10 kHz
[b] At 10 MHz

proportional to the square of the angular sound frequency, ω, and it varies with temperature. The dynamic and the bulk viscosities are included by Levy et al. (2001), as well as by Zaidni and Benjelloun (2021).

$$\alpha_s = \frac{2}{3}\frac{\eta\omega^2}{\rho c^3} \tag{2.16}$$

where η is the dynamic viscosity (Pa s), ω is the angular sound frequency (s^{-1}), ρ is the density (kg/m^3), and c is the sound speed (m/s).

The viscous sound attenuation values (Table 2.7, Sect. 2.3.2) are much higher for highly viscous glycerol than for water. The sound absorption is more pronounced at 10 MHz, than at 10 kHz. For increasing temperature, the sound absorption decreases in fresh water and increases in sea water (Brennan 2017).

The viscous and thermal thickness layers, δ_n and δ_T (μm), for water, glycerol and kerosene were calculated using Eqs. (2.14) and (2.15) (each liquid represents a homogenous phase only), Table 2.8. The viscous layers are thicker, therefore of greater significance, than the thermal layers. The calculated values are inversely proportional to the sound frequency.

2.5.5 *The Sediment Transport Theory*

There are four types of flow regimes for liquid and particles. These are the stationary bed, the sliding bed, the saltating particles, and the homogeneous flow (Rabenjafimanantsoa 2007). The last type is desired for ultrasonic measurements, in order to represent the average concentration in the cell. The homogenous flow was developed in the liquid-particle flow loop of the circular pipe's cross-section, for flow speeds exceeding 0.6 m/s (Rabenjafimanantsoa 2007). This is according to the observed flow homogeneity in the tests for the mixer speeds in water of 1.5–2 m/s (Sect. 2.5.6).

According to Nielsen (1992), particles are lifted from a sediment bed at a cell's bottom and get suspended in fluid, when the fluid's turbulent shear force exceeds the particle's gravity force, as in the presence of turbulent eddies. The particles are captured and partially trapped in vortexes (Rabenjafimanantsoa 2007; Nielsen 1992). The vortex trapping is the most important suspension maintaining mechanism. The vortex lifting mechanism is also involved in the transport of particles, by the means of the waves on fluid surfaces. The vortex trapping mechanism may affect measurement of the sound attenuation of suspensions, due to the inertia of the particles trapped in the vortexes.

2.5.6 Flow Regimes

The Reynolds numbers for all the mixer speeds in the performed tests were above 2,000. They indicate on the flow in high shear and turbulence. The flows of water for the mixer speeds of 1.56 and 1.61 m/s were steady, having a slightly wavy surface. Generally, no air bubbles were formed. For the speed of 1.71 m/s, the flow remained steady and no air bubbles were formed, however, the water surface was more deflected. For the last two mixer speeds of 1.84 and 1.98 m/s, the water flows looked turbulent. For the speed of 1.84 m/s, the water surface was quite wavy and very small air bubbles were occasionally formed, which flowed along the mainstream and the sound beam's path. For the last mixer speed of 1.98 m/s, the water surface was strongly wavy and more air bubbles were formed, including larger ones, sometimes flowing along the sound beam's path.

The flow regimes for the mixtures of water with glass particles were similar as for water. However, the mixtures were white and non-transparent; therefore, it was not possible to observe the entrained air bubbles. Nevertheless, the fluctuation of the fluid's surface was a good indication of the flow turbulence. The flow had the highest turbulence intensity for the mixer speeds of 1.84 and 1.98 m/s. No turbulent effect is expected to be reflected in the measurements for the mixer speeds up to 1.71 m/s. The water-glass bead mixtures in the test were well-mixed at all the speeds (Figs. 2.32 and 2.33).

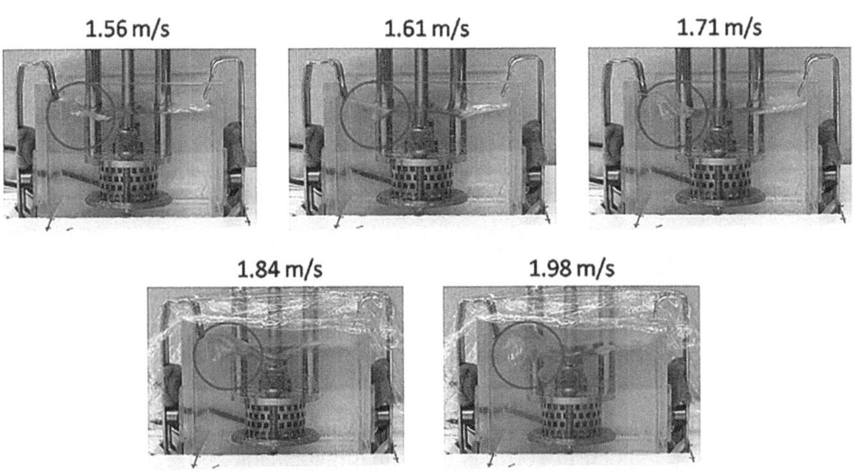

Fig. 2.32 The flow regimes for the tests with water, for the mixer speeds ranging from 1.56 to 1.98 m/s. (B.M. Wrobel, R.W. Time, *Applied Acoustics* 73 (2012) 117–131).

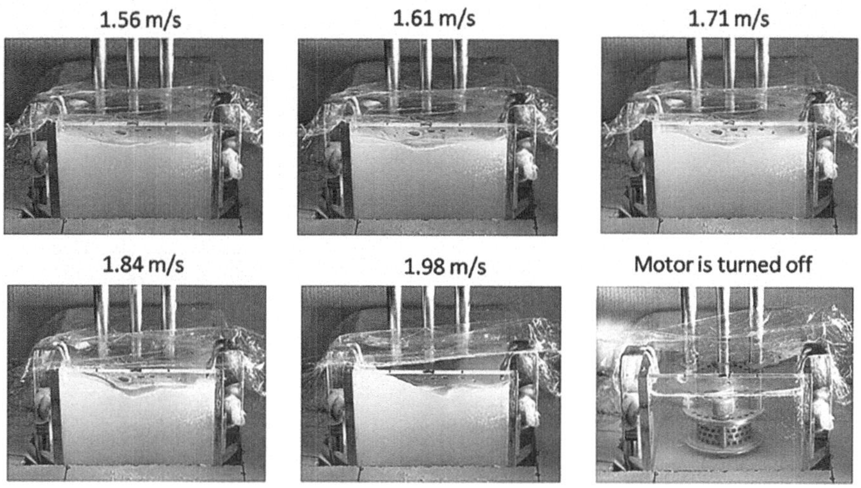

Fig. 2.33 The test flow regimes for the water-glass particle suspensions (the particle mass concentration of $1.78\%_{wt}$), for the mixer speeds ranging from 1.56 to 1.98 m/s. (B.M. Wrobel, R.W. Time, *Applied Acoustics* 73 (2012) 117–131).

2.5.7 *Sound Speed Dependent on the Particle Concentration and Flow Speed*

2.5.7.1 The Sound Speed Determination

Based on the total time of sound passing through a test cell, Fig. 2.34, the sound speed in fluid can be calculated using Eqs. (2.17) and (2.18):

$$\Delta T = 2\frac{L_1}{c_1} + \frac{L_2}{c_2} \tag{2.17}$$

$$c_2 = \frac{L_2}{\Delta T + C^* - 2\frac{L_1}{c_1}} \tag{2.18}$$

where:

ΔT—the time difference between the ultrasound incident pulse and the pulse received after passing through a square-shaped cell filled with fluid (s)
L_1 and L_2—the widths of the cell's wall and the sample of fluid (m)
c_1 and c_2—the sound speeds in the cell's wall and the sample of fluid (m/s)
C^*—the fixed calibration constant for the square signal, due to its shift to the right, $C^* = 4.97 \times 10^{-5}$ s ($C^* = 0$ for the spike pulse)

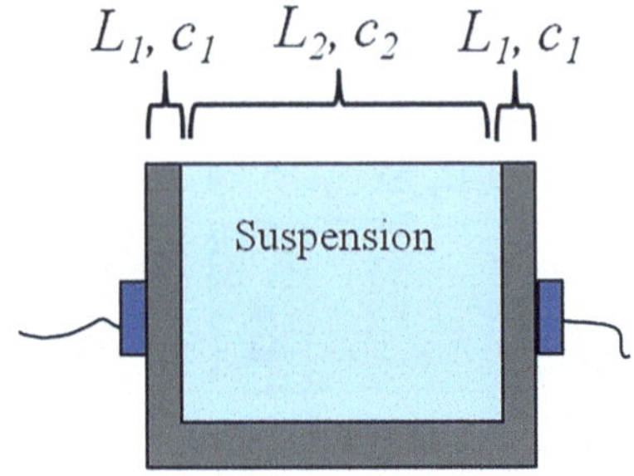

Fig. 2.34 The sound speed measurement of a suspension.

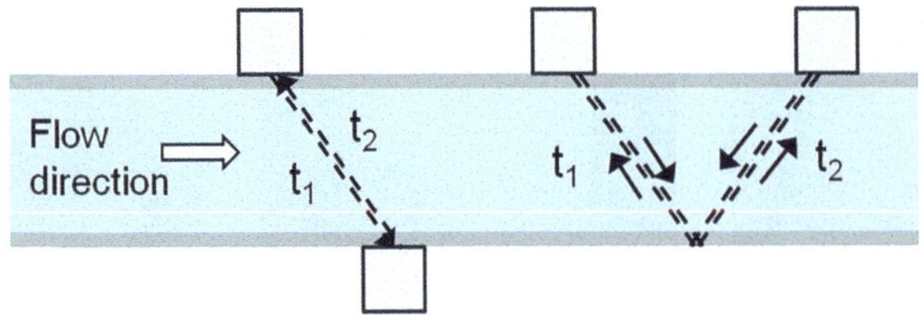

Fig. 2.35 An illustration showing the ultrasonic transducer configurations for the flow velocity measurement using the transit-time difference method.

2.5.7.2 The Flow Velocity Determination Using the Transit-Time Difference

According to the transit-time difference (TTD) method, the velocity of sound waves traveling in fluid is directly influenced by the velocity of that fluid. At no flow conditions, both sensors can receive the transmitted ultrasonic waves at the same time, meaning without any transit-time difference delay. At flow conditions, the ultrasound waves traveling "with" and "against" the flow directions would require different lengths of time (that depend on the flow speed) to reach the other sensor (Fig. 2.35). The measured transit-time difference is directly proportional to the flow velocity, u (m/s), Endress+Hauser (2006):

$$u = K \cdot \frac{t_1 - t_2}{t_1 \cdot t_2} \tag{2.19}$$

where:

t_1—the transit-time for the sound traveling "with the flow" (s)
t_2—the transit-time for the sound traveling "against the flow" (s)
K—the function of the length of the acoustic path, the ratio between radial and axial sensor distances, the velocity distribution and the flow channel's cross-section (m)

2.5.7.3 The Measurement in the "With the Flow" Direction

The measured square signals acquired "with the flow" for the mixer speeds of 1.56–1.98 m/s and 0.51%$_{wt}$ particle concentration are presented in Fig. 2.36. The phase shift is the measure of the sound speed. This relation is according to the superposition principle. It can be observed that as the flow speed increases, less time is needed to acquire the signal. This is performed by comparing the positions of the signals' peaks. For the sound traveling "against the flow", its speed would be reduced. In order to determine the average flow velocity using the transit-time difference method, both measurements in the "with the flow" and "against the flow" directions need to be performed using two transducers transmitting and receiving the ultrasonic waves at the same time.

2.5.7.4 Comments

In the case that an experimental setup does not allow for performing such tests, a solution is to switch transducers' connectors manually for the measurements in both flow directions; however, in this case quickly rising temperature of a system could introduce large measurement inaccuracies.

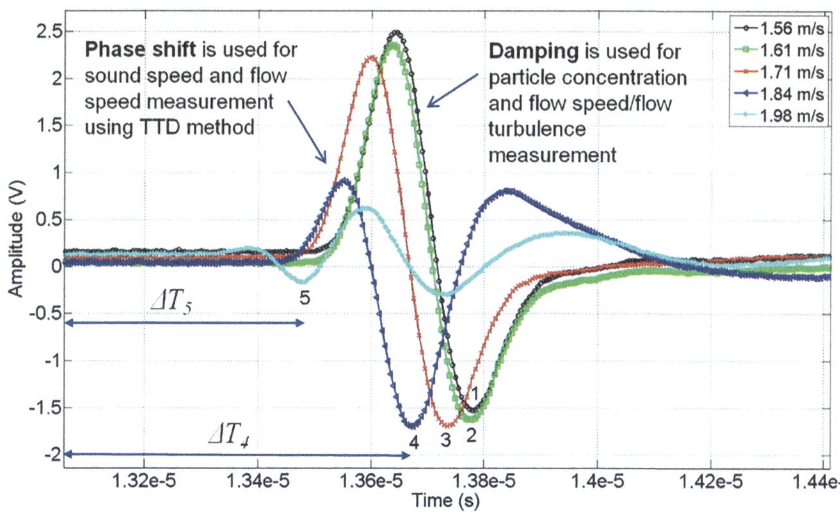

Fig. 2.36 The acquired square pulses traveling "with the flow," for the 0.51%$_{wt}$ water-particle suspension, for the mixer speeds of 1.56–1.98 m/s. (Upgraded, originally from B.M. Wrobel, R.W. Time, *Applied Acoustics* 73 (2012) 117–131).

Ultrasonic flowmeters using the transit-time method eliminate the effect of the sound wave velocity change due to changing fluid temperature, by the simultaneous measurement with and against the flow directions (Endress+Hauser 2006).

2.5.7.5 The Experimental Results Verification by Using Square and Spike Pulses as Two Independent Methods

The sound speed measurement of a flowing suspension was performed using two independent ultrasonic methods: the spike pulse from the pulser-receiver (500 Hz, 100 µJ) and the square pulse from the waveform generator (10 kHz, 5 V), for the results verification, Figs. 2.37 and 2.38. The data acquired using these two methods correlate well with each other.

The sound speed values for water with the glass particles (ϕ = 20–40 µm, ρ = 2.47 g/cm^3) follow the sound speed values for water closely for varying water temperature. The effective sound speed in a flowing medium should generally increase, when the sound wave travels in the direction of flow and can be thought of as the superposition principle. However, the sound speed values for the water-glass particle suspensions for the fractions higher than 1%$_{wt}$ were reduced, by the increased flow speed and flow turbulence intensity even by 5 m/s. This effect gets stronger for the increasing particle fraction (based on the measurement results using the pulser-receiver).

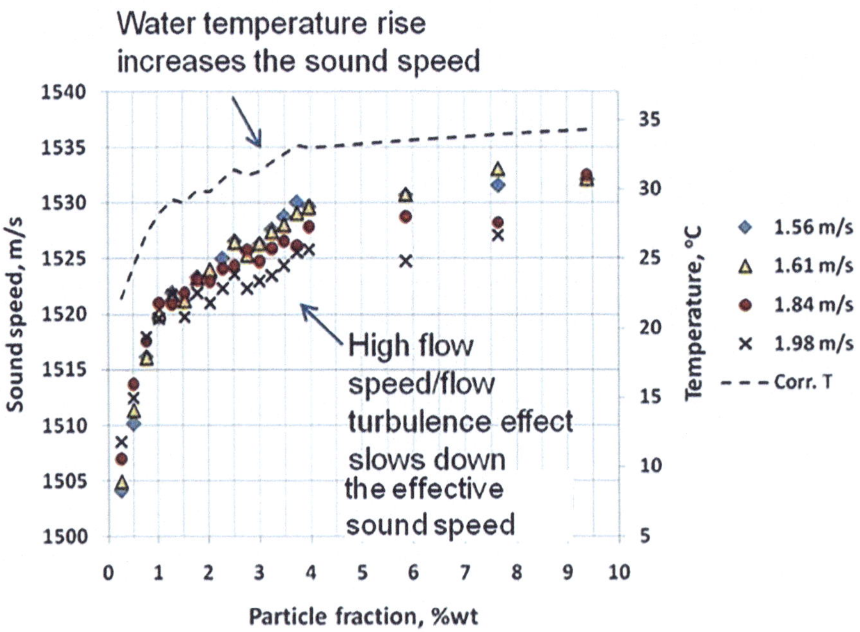

Fig. 2.37 The measured sound speed of the water-glass particle mixtures of the particle concentrations up to 9.38%$_{wt}$, for the mixer speeds of 1.56–1.98 m/s, using the 500 Hz spike pulse from the pulser-receiver. (B.M. Wrobel, R.W. Time, *Applied Acoustics* 73 (2012) 117–131).

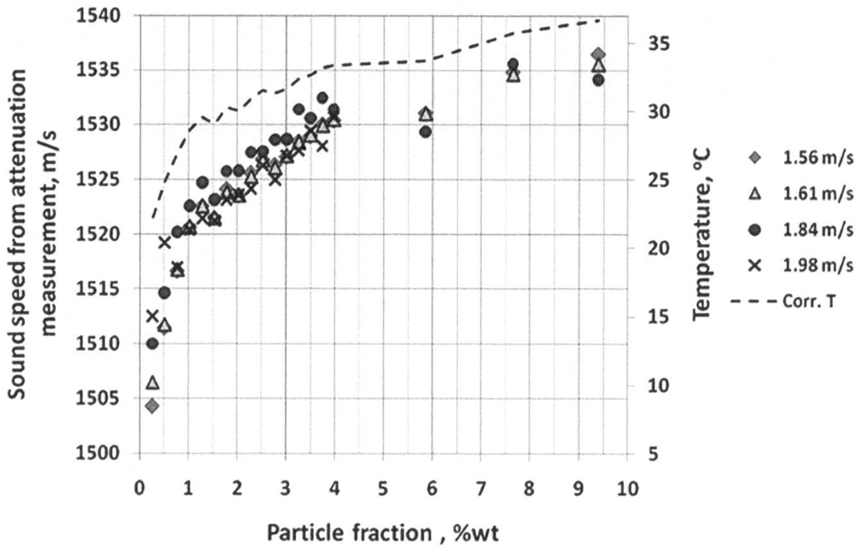

Fig. 2.38 The measured sound speed of the water-glass particle mixtures of the particle concentrations up to 9.38$\%_{wt}$, for the mixer speeds of 1.56–1.98 m/s, using the 10 kHz square pulse from the waveform generator. (B.M. Wrobel, R.W. Time, *Applied Acoustics* 73 (2012) 117–131).

Therefore, the reduced sound speed values for the increasing flow speed values can be explained by slowing down the sound by the flow turbulence (the sound energy and momentum were temporarily captured and redistributed into the turbulent flow vortexes).

2.5.7.6 The Sound Speed Measurement Uncertainty Factors

The measured sound speed values of the suspensions are influenced by several effects, from which three of them can be observed in Figs. 2.37 and 2.38:

1. The sound speed profile corresponds to the water temperature profile (Wrobel and Time 2012).
2. The sound speed values increase with the increasing flow speed values, for the particle fractions up to 1$\%_{wt.}$ Two phenomena exist here: the effective sound speed in the water-glass bead suspension increased, due to the higher sound speed value in glass (4540 m/s) than in water (1500 m/s), and there is the sound and flow superposition, due to traveling in the same direction.
3. The sound speed values get flattened for the particle fractions larger than 1$\%_{wt}$. This can be explained by the presence of the flow turbulence (the sound energy and momentum were temporarily captured and redistributed into the turbulent flow vortexes). It can be noted that the discrepancy between the data points corresponding to increasing particle fraction grows at the higher mixer speeds (referring to the spike pulse results in Fig. 2.37). This indicates the additional sound speed "slowing down" effect, due to the high flow speeds and high particle fractions.

Factors that impact the measured sound speed values of the suspensions are given below:

- The particle concentration, C_p (Case 1).
- The flow speed, v and the flow turbulence intensity, $\varepsilon(Re)$ (Case 2).
- The air bubble fraction, C_b (Case 3).
- Structural effects, like particle clustering (Case 4).
- The temperature of water, T_{water} (Case 5), and the temperature of particles, $T_{particles}$, if they are manufactured from a temperature sensitive material; neglected here due to low fractions.
- The temperature of the walls of the test cell for external measurement, T_{wall} (Case 6).

Some of these factors contributed to the increased sound speed values and some to their reduction, Fig. 2.39. The analysis was performed, based on the experimental results (Cases 1–2, 5–6), literature findings (Case 3) and a hypothesis (Case 4). The hypothesis states that the effect of particle clustering may reduce the sound speed, due to the lower "effective" particle fraction along the sound wave's path (Sect. 2.5.8). The estimation of impacts of the factors having influence on the effective sound speed of the tested suspensions is in Table 2.9.

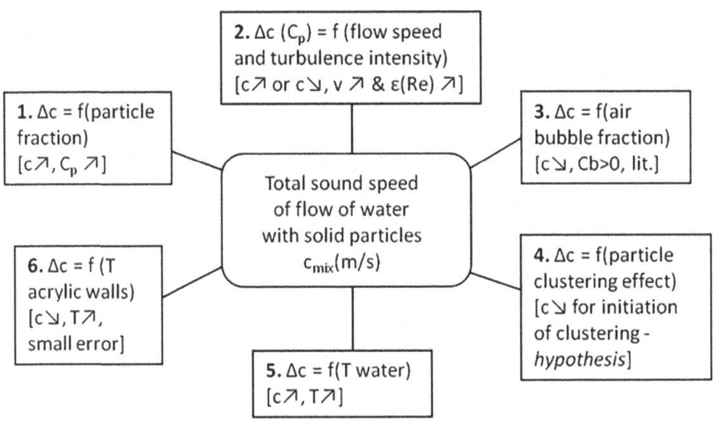

Fig. 2.39 Factors influencing the total sound speed of the low concentration water-glass particle mixtures ($<10\%_{wt}$), in high-shear flow. (B.M. Wrobel, R.W. Time, *Applied Acoustics* 73 (2012) 117–131).

Table 2.9 The estimated impacts of the factors that influence on the measured effective sound speed of the tested suspensions.

Case	Factor	Impact
1	Particle fraction	0.51 m/s per $1\%_{wt}$
2	Flow speed/turbulence intensity	1.54 m/s or larger per 1 m/s of flow speed
3	Individual air bubbles in flow	None in the performed test
4	Particle clustering	Unknown
5	Water temperature	2.5 m/s per 1 °C
6	Acrylic walls temperature	0.28 m/s per 1 °C (for calibrating out)

2.5.7.7 The Sound Speed Measurement for Liquid-Particle Flow

In order to answer the question about the effect of the particles only on the measured effective sound speed, other tests were performed. The tests were performed for three cases: water at static conditions, water flow for the mixer speeds of 1.56 and 1.84 m/s, and the water-glass particle flow for the same speeds, for the 3.02 and 4.28$\%_{wt}$ particle concentrations, in the temperature range of 20–36 °C.

The results in Fig. 2.40 show that the sound speed values increase with the increasing particle fraction. This is in contradiction with the phenomenological theory, Harker and Temple (1988) and the coupled phase theory (Gibson and Toksöz 1989, cited by Harker and Temple 1988). According to the sound speed results obtained by these authors for the suspension of the kaolin particles with the radius of 0.55 μm, the sound speed values decrease with the increasing particle fraction, in the low concentration range, up to 0.25$\%_{vol}$, and then they increase for the higher fractions, up to 0.5$\%_{vol}$. The most probable reason for this contradiction is that the effective compressibility may be overestimated (the predicted values are too high). In other words, there is too small gain of the suspension stiffness by the increasing particle fraction, compared to the values calculated in Tables 2.10 and 2.11. This simple analysis indicates that the effective sound speed of a suspension may increase with its increasing particle fraction, Fig. 2.41b, and not decrease, as predicted by the equations of Harker and Temple, and Urick, Eqs. (2.9)–(2.11).

On the other hand, the experimental results in Fig. 2.40 comply with the multiple scattering theory by Schwartz and Plona (1984), cited by Harker and Temple (1988), where the sound speed values of the kaolin particle suspension increase with the particle fraction within the whole concentration range. According to this theory, the sound intensity decreases with the squared number of particles. This can be interpreted that either the effective stiffness, or the effective density of a suspension, attenuates the sound more, or both effects influence it to a certain degree, Fig. 2.41a.

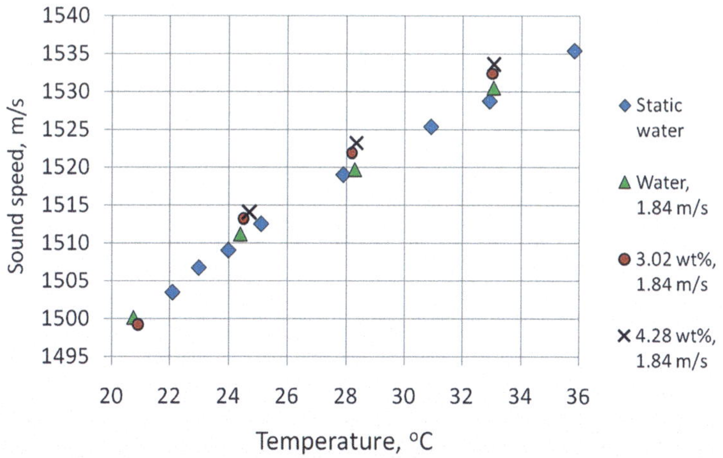

Fig. 2.40 The measured sound speed of water and water-glass particle flows, for the mixer speed of 1.84 m/s, dependent on temperature. (B.M. Wrobel, R.W. Time, *Applied Acoustics* 73 (2012) 117–131).

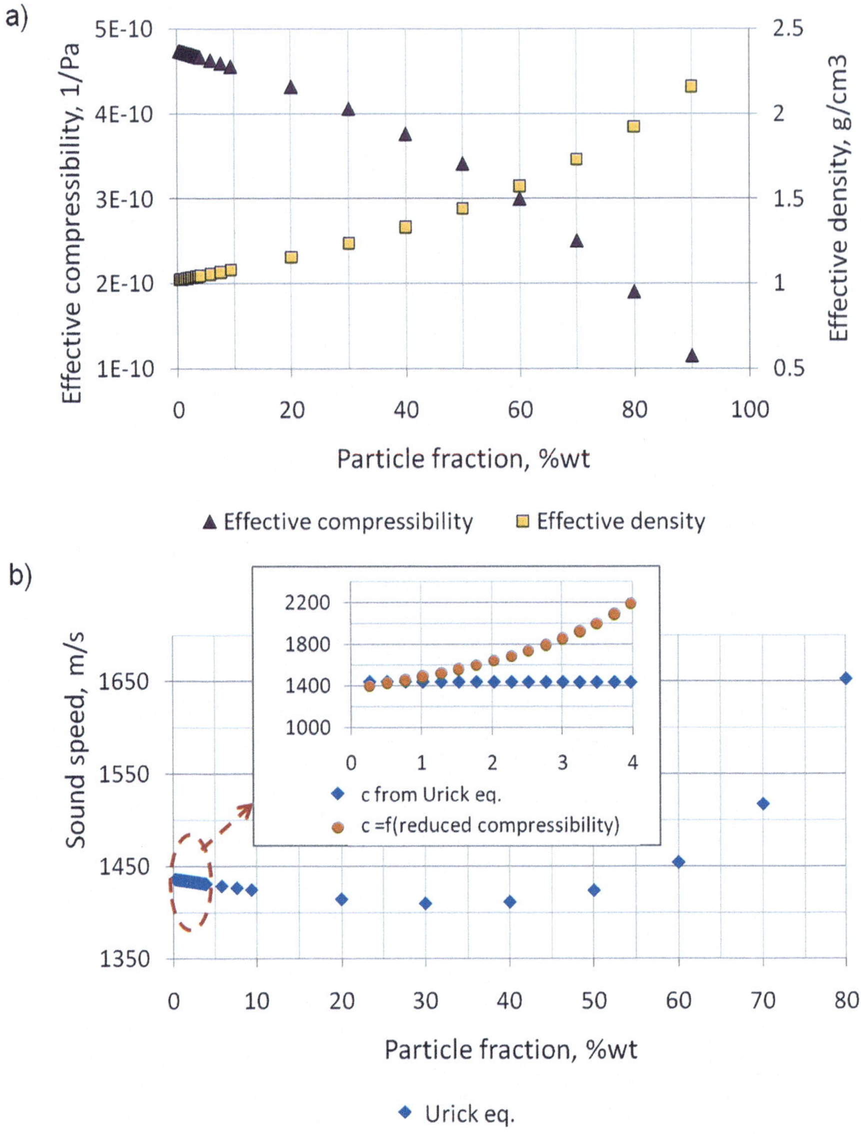

Fig. 2.41 (**a**) The theoretical effective compressibility and the effective density of a water-glass particle suspension, dependent on the mass fraction, according to Urick's equations, Eqs. (2.10) and (2.11). (**b**) The increasing sound speed values in the low concentration region are caused by the reduced compressibility values (Table 2.11).

Taking into consideration the calculations performed in Tables 2.10 and 2.11, it can be concluded that the effective compressibility of a suspension plays more important role, than the effective density. This leads to the increasing effective sound speed values, with the increasing particle concentrations.

Table 2.10 The calculated sound speed for water-glass particle suspensions, based on the equations of Harker and Temple, and Urick, Eqs. (2.9)–(2.11).

Particle mass fraction ($\%_{wt}$)	Effective density (g/cm^3)	Effective compressibility (Pa^{-1})	Sound speed (m/s)
0.26	1.022	4.745×10^{-10}	1436.30
0.51	1.023	4.740×10^{-10}	1435.96
0.77	1.025	4.735×10^{-10}	1435.61
1.02	1.026	4.731×10^{-10}	1435.28
1.28	1.028	4.726×10^{-10}	1434.93
1.53	1.029	4.721×10^{-10}	1434.60
1.78	1.031	4.716×10^{-10}	1434.27

Table 2.11 The calculated sound speed for water-glass particle suspensions, based on the equations of Harker and Temple, and Urick, Eqs. (2.9) and (2.10), with hypothetical values for the effective compressibility.

Particle mass fraction ($\%_{wt}$)	Effective density (g/cm^3)	Effective compressibility[a] (Pa^{-1})	Sound speed (m/s)
0.26	1.022	5.0×10^{-10}	1399.21
0.51	1.023	4.8×10^{-10}	1427.01
0.77	1.025	4.6×10^{-10}	1456.59
1.02	1.026	4.4×10^{-10}	1488.23
1.28	1.028	4.2×10^{-10}	1522.08
1.53	1.029	4.0×10^{-10}	1558.51
1.78	1.031	3.8×10^{-10}	1597.82

[a] The hypothetical values

2.5.8 *Sound Attenuation Dependent on the Particle Concentration and Flow Speed, with Data Science*

There are three important phenomena that need to be accounted for, when interpreting the attenuation spectra of suspensions):

1. A type of particles, which defines the type of dominating sound loss. For example, rigid and heavy particles will produce different shapes of the attenuation spectra than soft particles, due to the acting of a different sound loss mechanism (Dukhin and Goetz 2002).
2. Particle volume fraction, which defines the free space between particles that demarcates the sound frequencies that are the most and the least exposed to their attenuation.
3. Particle structural interactions, which can influence the "effective" particle size for the particles forming larger conglomerates at certain conditions (Fig. 2.42). On top of that there is the fourth flow-related phenomenon, which influences the attenuation of the frequency spectra for flowing suspensions:
4. Flow speed and flow turbulence.

Fig. 2.42 The floating clusters of glass particles initiated for the particle mass fraction of $3.74\%_{wt}$ (the appearance after turning off the mixer). (B.M. Wrobel, R.W. Time, *Applied Acoustics* 73 (2012) 117–131).

Regarding the first effect, the attenuation spectra for the glass beads ($\phi = 20$–40 µm, $\rho = 2.47$ g/cm^3), Figs. 2.43 and 2.44, resemble the spectrum obtained by Dukhin and Goetz (2002) for the quartz particles of the size of 120 µm in water. This spectrum, with a characteristic peak of the maximum sound attenuation, is obtained for rigid, heavy particles in water.

Regarding the second effect, a frequency at which a sound wave reaches the maximum attenuation is called the critical frequency, ω_{cr}. It is related to an instance of the particle size or the interparticle distance, from which the latter depends on the particle volume fraction, according to the viscous loss theory. Based on the test performed for the mixer speed of 1.56 m/s, the measured critical frequency is 8.9 MHz for the particle fraction of $2.03\%_{wt}$ ($0.85\%_{vol}$), and the calculated particle size is 0.12 µm. This size is much smaller than the actual particle sizes of 20–40 µm, which may suggest that not the high-volume fraction, but the high-shear flow effects strongly reduced the viscous layer thickness and promoted the particle interactions.

Regarding the third effect, it can be noticed in Fig. 2.43 that the critical frequency slightly increases for the fractions up to $3.01\%_{wt}$, from approximately 8.3 to 8.9 MHz, and then it is significantly reduced to 8 MHz for the fraction of $5.85\%_{wt}$. This increase of the critical frequency can be associated with the increasing particle fraction, as also obtained by Dukhin and Goetz (1996). Whereas the drop in the critical frequency could be explained by the particle clustering effect initiated at the particle concentration of $3.74\%_{wt}$ (as it was observed during the test, Fig. 2.42). This would result in the critical frequency reduction for a new, larger "particle size" created from the agglomerated particles.

Regarding the fourth effect, the sound attenuation in the water-particle mixture depends on the flow speed, Fig. 2.44. It is greater for the increasing flow speed values. This effect is significant for the mixer speeds of 1.84 and 1.98 m/s. The sound

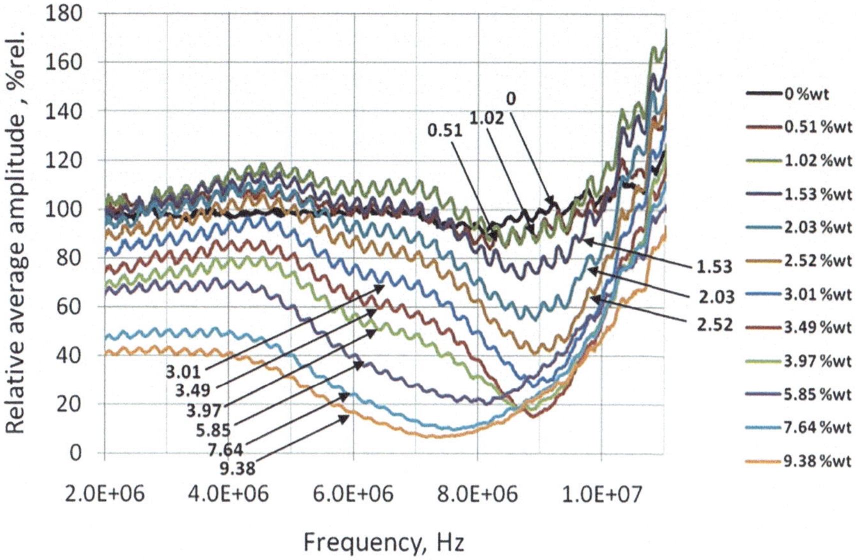

Fig. 2.43 The measured relative sound attenuation of ultrasonic square pulses, for the flowing water-particle suspensions of the particle mass fractions of 0–9.38%$_{wt}$, for the mixer speed of 1.56 m/s. The particle clustering effect is visible through the frequency shift starting at 5.85%$_{wt}$. (B.M. Wrobel, R.W. Time, *Applied Acoustics* 73 (2012) 117–131).

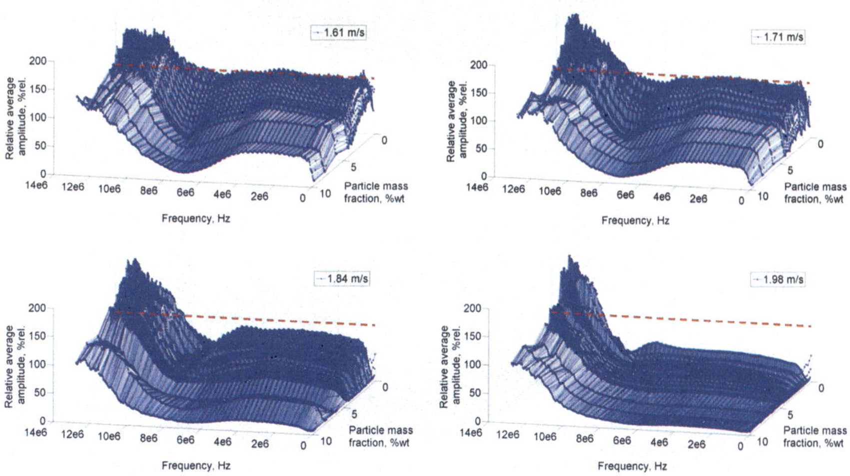

Fig. 2.44 The measured relative average sound attenuation of ultrasonic square pulses, for the flowing water-particle suspensions of the particle mass fractions of 0–9.35%$_{wt}$, for the mixer speeds of 1.61–1.98 m/s, and for the sound frequencies available for measurement of 1–14 MHz. (B.M. Wrobel, R.W. Time, *Applied Acoustics* 73 (2012) 117–131).

attenuation takes place within the whole useful sound absorption range of 2–12 MHz, but the highest attenuation is around the critical frequency of 8–9 MHz. It can be concluded that the sound attenuation depends on the flow speed, most likely through the flow turbulence intensity. This effect is due to the sound absorption and sound scattering by the internal turbulent flow structures.

The average standard deviation values in Fig. 2.45 became increasingly unstable for all the mixer speeds. The values increase up to the particle concentration of 3.74%$_{wt}$. The interpretation of this effect is the following: the signal fluctuations increase with the increasing particle fractions, due to the particle inertia force becoming more powerful, affecting the flow stability and therefore the instantaneous ultrasonic signals. The standard deviation values for the higher fractions are approximately constant, however, only a few points were measured here. The extra disturbances for the mixer speed of 1.84 m/s are related to the high flow turbulence intensity.

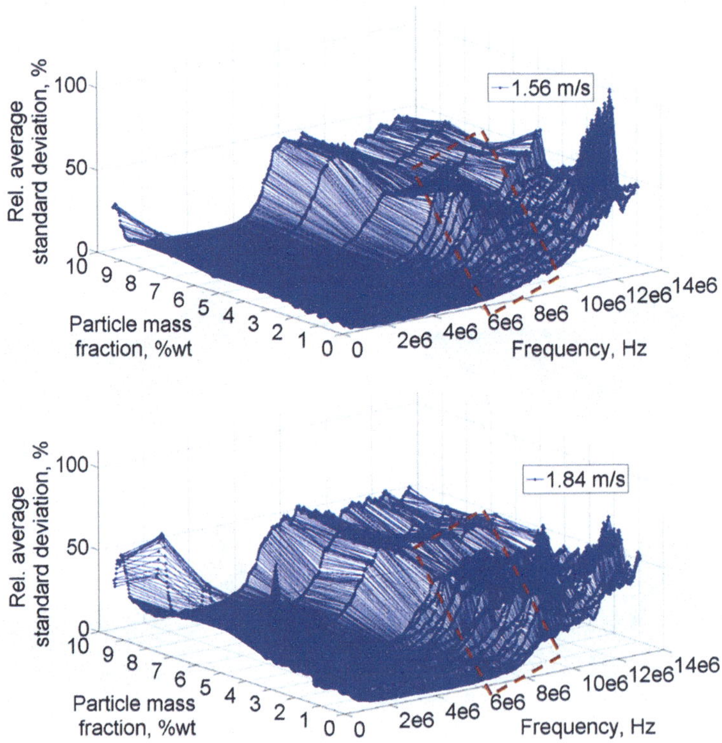

Fig. 2.45 The measured relative average standard deviation of ultrasonic square pulses, for the flowing water-particle suspensions of the mass fractions of 0–9.35%$_{wt}$, for the mixer speeds of 1.56 and 1.84 m/s, and for the sound frequencies available for measurement of 1–14 MHz. (B.M. Wrobel, R.W. Time, *Applied Acoustics* 73 (2012) 117–131).

2.5.8.1 The Evaluation of the Sound Measurement Accuracy

The coefficient of the ultrasonic absorption in pure water is directly proportional to the ultrasonic pulse frequency, according to Stokes' equation, Eq. (2.16), Sect. 2.5.4. To verify the accuracy of the spectroscopy method, the sound attenuation coefficient was calculated, for the test cases of water at static conditions, and the concentrated suspensions for the mixer speeds of 1.56 and 1.98 m/s, using the equation from Dukhin et al. (2005):

$$\alpha\left[\mathrm{dB/cm/MHz}\right]=\frac{10}{f\left[\mathrm{MHz}\right]L\left[\mathrm{cm}\right]}\log\frac{I_{\mathrm{in}}}{I_{\mathrm{out}}} \tag{2.20}$$

where I_{in} is the reference intensity of the ultrasound pulse acquired directly from a signal generator (corresponding to the measurement in Fig. 2.46, the upper, smooth signal), I_{out} is the pulse intensity after propagation through transducers and a sample (corresponding to the measurement in Fig. 2.46, the lower, wavy signal), f is the pulse frequency (MHz), and L is the sample width (cm). The intensity level unit of the logarithmic ratio, I_{in}/I_{out}, is in dB.

The test values for the case of water at static conditions, for example, α(10 MHz) = 0.1 dB/cm/MHz are a little bit higher, than the reference values for distilled water obtained by Dukhin and Goetz (2002), α(10 MHz) = 0.025 dB/cm/MHz and α(52 MHz) = 0.1 dB/cm/MHz. The sound attenuation in the cell's walls influenced (increased) the total measured attenuation, because water and the walls were measured together.

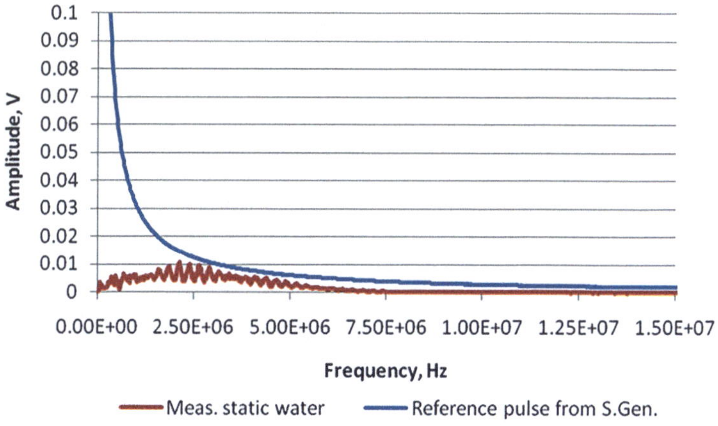

Fig. 2.46 The acquired spectra of the incident square pulse from the signal generator and the pulse for water at static conditions (the wavy signal), using the through-transmission method.

The shapes of the spectra for particles, for the mixer speed of 1.56 m/s, resemble the shapes obtained by Dukhin and Goetz (2002) for the 120 μm quartz particles in water. Here, the maximum attenuation is at 9 MHz, having the value of 1.5 dB/cm/MHz for the particle volume fraction of $3\%_{vol}$, and it is 2.5 dB/cm/MHz for the particle volume fraction of $5\%_{vol}$.

For comparison, the experimental results give the maximum sound attenuation of 0.18 dB/cm/MHz, for the 20–40 μm, $2.03\%_{wt}$ ($0.85\%_{vol}$) glass particles at 8.9 MHz, and the maximum sound attenuation of 0.23 dB/cm/MHz, for the particle volume fraction of $3.97\%_{wt}$ ($1.7\%_{vol}$) at 8.5 MHz.

The comparison using proportional relationships leads to the conclusion that the test data are within the proper sound attenuation range. When it comes to the spectra obtained for the mixer speed of 1.98 m/s, they are modified in shape, due to the turbulent flow speed. This is according to the hypothesis, stating that fluid flow speed would influence the sound attenuation measurement.

The calculated sound attenuation coefficient value for water at static conditions indicates good accuracy of the spectroscopy method within the frequency range of 4–12 MHz, with the used experimental setup. The characteristics of the ultrasonic transducer determines its broadband measurement limit, as well as the quality of the incident wave. Within the 4–12 MHz frequency region, the measured sound attenuation values increase with the frequency, Figs. 2.47, 2.48, and 2.49.

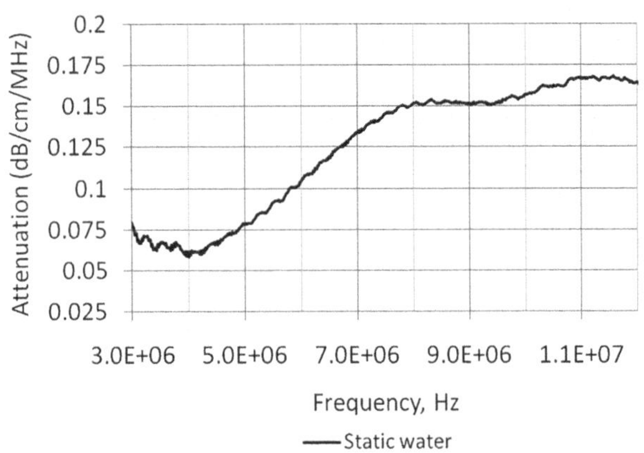

Fig. 2.47 The acquired attenuation frequency spectrum for water at static conditions, corresponding to the data in Fig. 2.46. (B.M. Wrobel, R.W. Time, *Applied Acoustics* 73 (2012) 117–131).

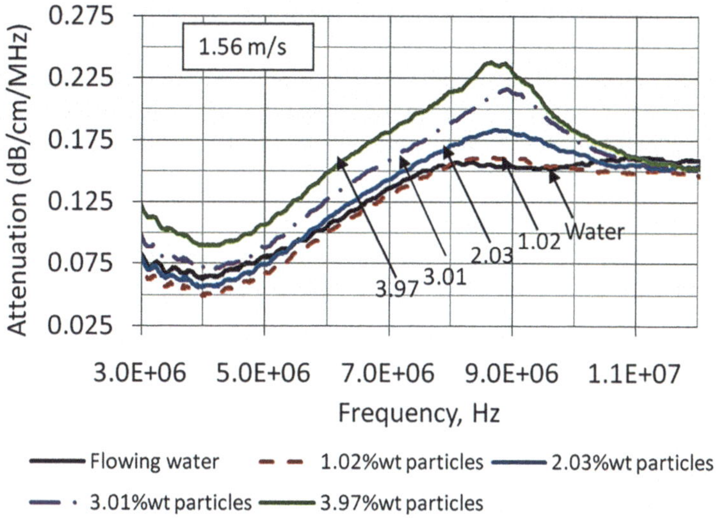

Fig. 2.48 The attenuation frequency spectra for the cases of water and the water-particle mixtures (1–4%$_{wt}$), for the mixer speed of 1.56 m/s. (B.M. Wrobel, R.W. Time, *Applied Acoustics* 73 (2012) 117–131).

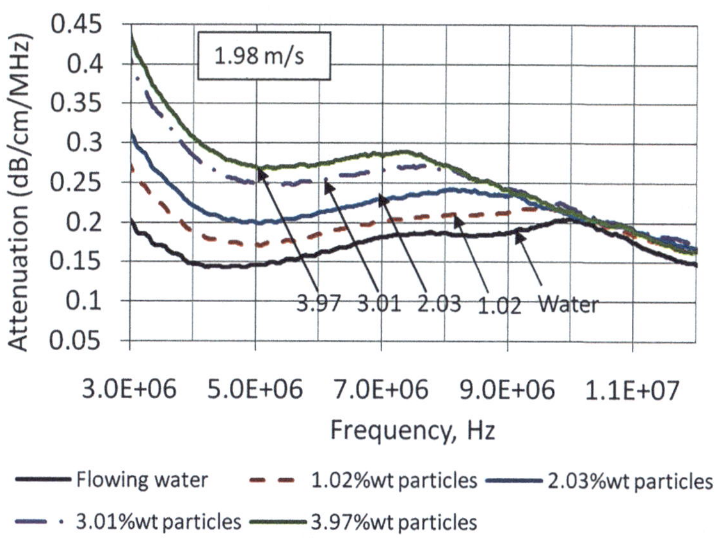

Fig. 2.49 The attenuation frequency spectra for the cases of water and the water-particle mixtures (1–4%$_{wt}$), for the mixer speed of 1.98 m/s. (B.M. Wrobel, R.W. Time, *Applied Acoustics* 73 (2012) 117–131).

2.5.8.2 The Sound Attenuation Measurement Uncertainty Factors

Factors that contribute to the overall uncertainty of the ultrasound attenuation measurement for water-particle flow are presented in Fig. 2.50. Based on the experimental results, effects that reduce the amplitude (increase the sound attenuation) are the following:

- The particle fraction, C_p (Case 1).
- The flow speed, v and the flow turbulence intensity, $\varepsilon(Re)$ (Case 2).
- The intrinsic losses in a continuous medium, such as water (Case 8).

Effects that increase the amplitude (decrease the sound attenuation) are the following:

- The air bubble fraction, C_b (Case 3) at frequencies higher than 12 MHz.
- The moment of initiation of particle clustering (Case 4), according to the hypothesis that binding of small particles into clusters will result in a smaller number of particles, along the sound wave's path, than for non-clustered ones. The transferred sound energy will increase for the clusters that are even of higher fraction than of the non-clusters, as it occurred in the test, Fig. 2.43.
- The increasing temperature of the acrylic walls, T_{wall} (Case 5) and water, T_{water} (Case 7). However, the latter effect was not verified by the temperature test.

The sound attenuation values for the suspensions (Case 6) depend on the reference amplitude for water (Case 8), which in turn is dependent on the medium temperature, due to the relative plotting, Figs. 2.43 and 2.50.

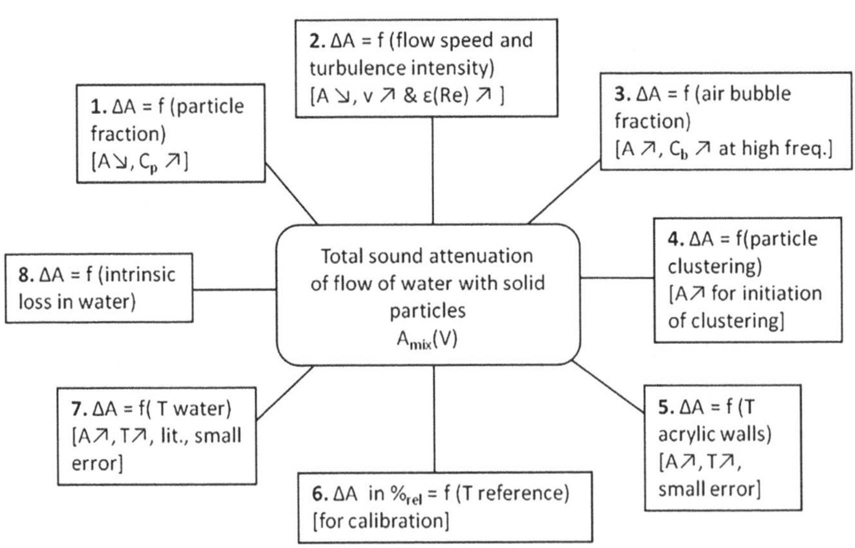

Fig. 2.50 Factors influencing the total sound attenuation of the low concentration water-glass particle mixtures (<10%$_{wt}$), in high-shear flow. (B.M. Wrobel, R.W. Time, *Applied Acoustics* 73 (2012) 117–131).

2.5.9 *Small Gas Bubbles in the Liquid-Particle Flow*

2.5.9.1 The Ultrasonic Nature of Air Bubbles

Individual air bubbles were occasionally flowing along the sound wave's path, especially for the high mixer speeds of 1.84 and 1.98 m/s, which could influence the accuracy of the sound attenuation measurement. Conclusions from the test are the following:

- The air bubbles produced visible noise in the form of a spike or a cloud in the time signal, Fig. 2.51a.
- The air bubbles amplified the sound frequencies higher than 12 MHz, however, generally they did not influence the sound frequencies lower than 12 MHz. The latter frequencies, Fig. 2.51b, are useful for testing of water with glass particles.
- The signals acquired for the particle suspension had better statistics (a high number of samples was acquired), compared to the signals that captured air bubbles, which appeared only sporadically.

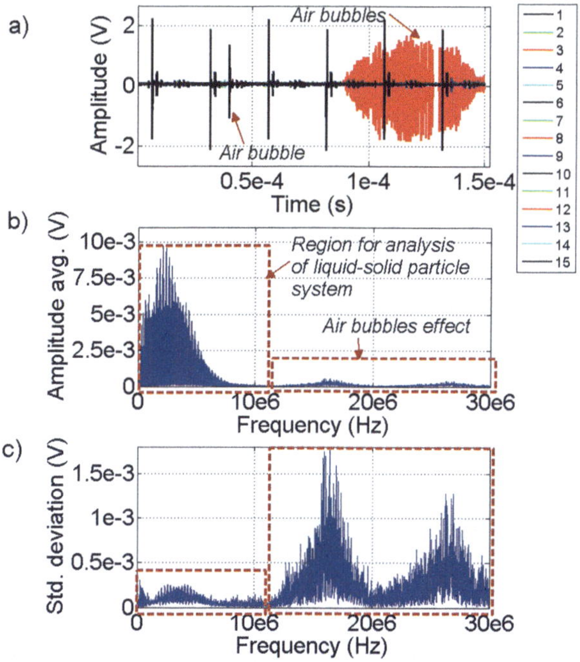

Fig. 2.51 (**a**) Fifteen acquired square signals, from which two of them "captured" air bubbles. (**b**) The frequency ranges for the water-glass particle system (up to 12 MHz) and for the air bubbles (higher than 12 MHz). (**c**) The corresponding average standard deviation plot. (B.M. Wrobel, R.W. Time, *Applied Acoustics* 73 (2012) 117–131).

- The flow turbulence intensity is reflected in the standard deviation plots, Fig. 2.51c. The frequencies higher than 8 MHz experienced large deviations, while the lower frequencies of 10 kHz–8 MHz were affected in the way that the average deviation level was not exceeded. The presence of air bubbles contributed to the high values of standard deviation, in addition to the turbulent movements of water and the particles.

The sound amplitude amplification by air bubbles within the high-frequency range, Fig. 2.51, could be explained by their vibrations in the sound field causing the sound resonance, when the ultrasound wave propagated through them. The sound is scattered by the air bubbles and the high sound frequencies are affected. The viscous and thermal sound absorptions by the bubbles are relatively less important; the low sound frequencies are generally not affected.

A conclusion is that the measurement error, due to the presence of individual air bubbles in flow is insignificant within the frequencies lower than 12 MHz (the useful measurement frequencies). However, signals with large noise from air bubbles should be discarded from measurement results, to ensure the highest measurement accuracy.

2.5.10 Conclusions

2.5.10.1 General Conclusions

- Low concentration systems of solid particles in liquid in high-shear flow can be studied and measured using the pulsed broadband ultrasonic attenuation spectroscopy and the sound speed techniques. The spectroscopy method is primary, while the sound speed method is complementary and verifying.
- There is potential for the simultaneous estimation of the average particle concentration and the flow speed in high shear, using a square pulse for the spectroscopy measurement, and the transit-time difference technique for the sound speed measurement. The sound attenuation depends on the particle concentration and flow speed parameters, while the sound phase shift (the sound speed) depends on the flow speed.
- The flow turbulence intensity in liquid and liquid-particle systems can be studied using a combination of the mentioned techniques, however, especially using the sound attenuation spectroscopy measurement.
- The calculated sound attenuation coefficient for water at static conditions indicates good accuracy of the spectroscopy method, within the frequency range of 4–12 MHz, with the utilized experimental setup. The experimental results from the attenuation test of the suspensions point to the useful measurement frequency range of 2–12 MHz (the noise region begins at higher frequencies and is characterized by the visible, distinctive signal's irregularities).

- The effects of particle clustering and individual air bubbles in liquid flow with solid particles can be studied using the spectroscopy method.
- Effects, such as the flow turbulence, the air bubble fraction, the particle clustering, and the temperature of a whole measurement system (for example, the system used in this project consisted of the flowing suspension of water with glass particles and the walls of the test cell, for the external type of measurement) need to be considered for the correct interpretation of the sound attenuation signals acquired for suspensions.
- An important discovery is that the superposition theory proposed by Dukhin and Goetz (1996, 2002, 2017), Eq. (2.13), could be extended to one more element, α_{turb}, which applies for turbulent flow applications. Therefore, this extra element is added to the total sound attenuation in colloids: $\alpha_{\mathrm{T}} = \alpha_{\mathrm{vis}} + \alpha_{\mathrm{th}} + \alpha_{\mathrm{st}} + \alpha_{\mathrm{sc}} + \alpha_{\mathrm{int}} + \alpha_{\mathrm{el}} + \alpha_{\mathrm{turb}}$, where α_{turb} is the sound loss mechanism present in turbulent flow, where turbulent structures like vortexes and eddies, attenuate the sound traveling through them. This sound loss is especially pronounced in the case of a liquid-particle suspension; however, it is also present for liquid flow. The levels of dominating sound loss mechanisms will be different for different applications.
- To account for the turbulence effects in the acoustic models for particulates, the turbulent flow regime and the related turbulent drag coefficient for high Reynolds numbers need to be considered (please refer to the work by Manninen and Taivassalo 1996).

2.5.10.2 Conclusions from the Sound Speed Measurement

- The effective sound speed in a flowing medium should generally increase, when the sound wave propagates in the direction of flow, and can be thought of as the superposition of the sound speed and flow speed. However, the sound speed in water with glass particles for the fractions higher than $1\%_{\mathrm{wt}}$ was reduced, by the increasing flow speed values (and the flow turbulence intensity), even up to 5 m/s. This effect gets stronger, as the particle fraction increases (based on the measurement results using the pulser-receiver).
- The values of the sound speed in water with particles follow the values of the sound speed in water closely, for the varying water temperature (Fig. 8, Wrobel and Time 2012). The particle presence increased the sound speed for the fixed flow speed (not highly turbulent) and constant temperature.
- This result complies with the results described by the multiple scattering theory by Schwartz and Plona (1984), cited by Harker and Temple (1988), but does not comply with the hydrodynamic theory, Harker and Temple (1988), and the phenomenological theory by Urick (1947), cited by Harker and Temple (1988). This may be explained by the effective compressibility overestimation (it shall be lower than predicted) by the hydrodynamic and the phenomenological theories, dominating the effective density in the low concentration region,

otherwise considered in the multiple scattering theory. The overestimation of the effective compressibility of the tested suspension could be calibrated for, by using a factor.

- To eliminate the temperature influence on the sound speed measurement, the sound speed needs to be measured simultaneously for the "with the flow" and "against the flow" directions. This is how it is performed in commercial ultrasonic flowmeters, using the transit-time difference method. The flow speed acquired in this way can be measured independently of temperature. Such settings were not possible in the test.
- The sound speed values in the acrylic wall decreased with increasing temperature by 2.86 m/s per 1 °C (Fig. 10, Wrobel and Time 2012). This caused a slight underestimation of the measured sound speed of the water with glass particle suspension. For thin walls, this effect is represented by a small fraction of the total sound speed, which can be calibrated for.

2.5.10.3 Conclusions from the Sound Attenuation Measurement

- In the sound attenuation tests, the signal amplitude within 2–12 MHz decreased for the increasing particle mass fractions up to $9.38\%_{wt}$. This relation was sustained for all the mixer speeds from 1.56 to 1.98 m/s. The signal amplitude decreased from 100% to approximately 10%, relatively to the reference signal for water at static conditions and the mixer speed of 1.56 m/s, and from 55% to approximately 3%, relatively to the reference signal and the mixer speed of 1.98 m/s.
- The comparison of the sound attenuation values for the suspensions, for the lowest mixer speed of 1.56 m/s, with the reference data obtained by Dukhin and Goetz (2002), brings a conclusion that the obtained sound attenuation accuracy is within the expected range. For example, for the $2.03\%_{wt}$ ($0.85\%_{vol}$) suspension, the maximum sound attenuation coefficient is 0.18 dB/cm/MHz, and for the $3.97\%_{wt}$ ($1.7\%_{vol}$) suspension, it is 0.23 dB/cm/MHz.
- For the cases of the flow tests of water and water with the particles suspensions, the sound amplitude decreased from $55\%_{rel}$ (0.18 dB/cm/MHz) for the particle concentration of $2.03\%_{wt}$ and the mixer speed of 1.56 m/s, to $19\%_{rel}$ (0.24 dB/cm/MHz) for the mixer speed of 1.98 m/s, at the example frequency of 8.8 MHz. This confirms the measurement sensitivity of ultrasounds to the flow turbulence intensity. According to Winkler (2010): "Only recently, some of ultrasonic Doppler flowmeters have been developed to work with clean liquids, reflecting off turbulent swirls."

- The standard deviation values of the sound attenuation spectroscopy increased for the increasing particle fractions up to $4\%_{wt}$, as the result of the enhancement of the flow turbulence intensity by the particle inertia force. In qualitative terms, the standard deviation increased from 15% for pure water to 55% for the $3.97\%_{wt}$ water-particle suspension, at the chosen frequency of 8.8 MHz.
- The sound loss mechanism, α_{turb}, due to the flow speed and flow turbulence intensity had significant influence on the ultrasonic measurements, and it needs to be accounted for in the acoustic theories to a higher degree.
- The results showed that sound attenuation for the liquid-glass particle suspension was basically not dependent on individual air bubbles that were sporadically present in the flow. This is because the air bubbles affected much higher sound frequencies than the suspension. However, according to the rule of thumb, it is best to avoid high amounts of air bubbles, because they will attenuate ultrasounds at all frequencies. The remedy in this case is to produce better measurement statistics for an air-free mixture. This means essentially, to acquire a signal for a long time or to acquire repeated time series.
- Generally, sound attenuation in water decreases with its increasing temperature (Brennan 2017). However, the temperature tests for water at static conditions did not result in any clear trends (Sect. 4.3 "Water Test," Wrobel and Time 2012). This was probably due to the insufficient sensitivity of the measurement system for the detection of the sound attenuation in water dependent on temperature. After all, the system consists of the cell's walls and the cell's volume. An issue with the signals' stability and the small amplitude occurred during that temperature test. Reasons for this could be: the unfavorable acoustic impedance of the transducer-water interface (less likely), the non-perfect alignment of the transmitter and receiver faces (less likely), a possible lack of the external pressing force on the transducers in that particular test (more likely), required higher sensitivity of the ultrasound system for the sound attenuation measurement dependent on the temperature of water (more likely). The latter can be fulfilled by direct, non-external measurement. According to the results obtained by Dukhin et al. (2005), the sound attenuation depends mostly on the composition and concentration of the dispersed phase of a colloid, being less sensitive to the temperature variations.
- The sound attenuation values in the acrylic wall decreased with the increasing temperature, with the rate of 6×10^{-5} V per 1 °C. This is approximately $1.2\%_{rel}$ of the average amplitude for the mixture (Sect. 4.3 "Acrylic Wall Test," Wrobel and Time 2012). This effect is small, however, it needs to be accounted for in the through-wall measurements. Additional care needs to be taken for sound resonances, since measurements in different test cells will be influenced by the resonance effects, created inside the walls and volume of test cells.

2.6 4D Finite Element Modeling of the Acoustic Fields in Liquid-Particle Systems

2.6.1 Subproject Objectives and Challenges

2.6.1.1 The Subproject Objectives

It is desired to understand the nature of sound waves propagation in media. The objective of this study is to calculate the acoustic fields in liquid and liquid-particle systems. In other words, the distribution of the sound wave pressure dependent on time, which is related to the sound attenuation in these systems. Simulations relate to some degree to the earlier described experimental results. The effects of different particle distributions and particle clustering on the sound pressure are analyzed.

2.6.1.2 Comments

Only partial verification of the experimental results is possible for modeling. This means that some of the sound loss mechanisms can be studied, namely the intrinsic loss that is related to the medium density, the scattering loss in terms of the reflection from liquid-particle boundaries, and the structural loss in terms of particle clustering. The viscous and flow speed losses (including the flow turbulence intensity) are neglected, although they play important roles in the total sound attenuation, as the experimental results have revealed. The thermal and electrokinetic losses are neglected in the planned simulations.

2.6.1.3 The Subproject Challenges

- Simulating the propagation of ultrasonic waves and the sound pressure distributions in water and water-glass particle systems.
- Comparing the simulation results with the experimental results.

2.6.2 Visualizing Sound Energy

The visualization of the sound energy in a liquid-particle system is illustrated in Fig. 2.52. The total attenuation of the sound beam that reaches a particle is composed of the sound scattering and sound absorption parts. Certainly, the received sound energy after traveling through the particle suspended in the liquid is diminished, by the sound absorption in the liquid, particle and cell's walls, for the external type of measurement. Additionally, the sound reflections within the walls and cell's volume will also diminish the sound energy.

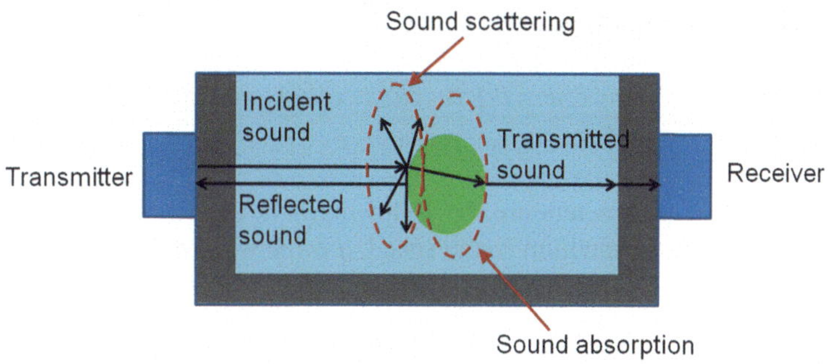

Fig. 2.52 The visualization of the sound attenuation in a liquid-particle system.

2.6.3 *Simulation Design*

2.6.3.1 The Incident Source Wave and Simulation Scenarios

4D finite element simulations (3D with the element of time) were performed using the acoustic module in COMSOL Multiphysics. The 2 MHz frequency damped sine wave was used as the incident source wave, Eq. (2.21). The wave equation, Eq. (2.22), was used for the determination of the wave propagation in time, and the wave pressure distribution in space. A layer of water was designed with the length of 28 mm and the thickness of 10 mm, containing the round quartz particles of the radius of 250 μm, evenly distributed in the layer.

$$y(t) = A \cdot e^{-\left(\frac{t-t_0}{\sigma}\right)^{10}} \cdot \sin(\omega_0 t) \tag{2.21}$$

Three scenarios were studied: (1) The case of pure water. (2) The case of water with uniformly distributed particles having the concentration of 0.066% (by area). (3) The case of water with the same number of particles, however, joined in clusters.

For Cases 2 and 3, two different particle distributions were studied. The absolute sound pressure, U_{sum}, along the beam's path and across the fluid, was used for comparison, at the fixed time step of 6.650×10^{-6} s. This is the time step before the wave is reflected from the water-air interface, where the reflection would occur at 13.33 wavelengths, at the time step of 6.665×10^{-6} s. Related calculations are the following: one wavelength travels in water for $\lambda = c/f = 1500/(2 \times 10^6) = 0.75$ mm, while in 10 mm there are 10/0.75 = 13.33 wavelengths. One wavelength travels for $T = 1/f = 5 \times 10^{-7}$ s, and 13.33 wavelengths travel in 6.665×10^{-6} s.

The incoming wave is generated by a transducer source located on the right-hand side of the layer. Other simulation parameters are the time stepping, for example, $1 \times 10^{-8} : 4 \times 10^{-8} : 1 \times 10^{-5}$ s, the triangular finite element type and the Neumann boundary condition (the natural boundary condition), $n \cdot (c \nabla u) = 0$, for the external boundaries.

2.6.3.2 The Wave Equation

$$f = \frac{1}{\rho_0 c^2} \cdot \frac{\partial^2 u}{\partial t^2} - \frac{\partial}{\partial x}\left(\frac{1}{\rho_0} \cdot \frac{\partial u}{\partial x}\right) - \frac{\partial}{\partial y}\left(\frac{1}{\rho_0} \cdot \frac{\partial u}{\partial y}\right) \tag{2.22}$$

The wave equation parameters are the following: f is the source wave, ρ_0 is the medium density, c is the medium sound speed, u is the wave pressure, t is the wave propagation time, x is the distance in the x-direction, y is the distance in the y-direction.

2.6.4 Results

The modeling results of acoustic fields in water and water-particle systems are presented in Figs. 2.53, 2.54, 2.55, 2.56, 2.57 and Table 2.12.

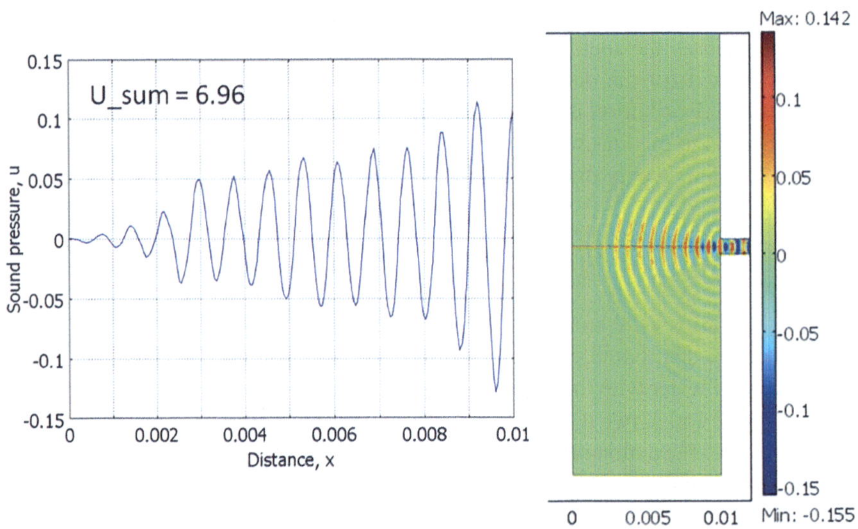

Fig. 2.53 On the RHS: the acoustic field in the layer of water. On the LHS: the cross-sectional plot of the sound pressure, along the longitudinal axis.

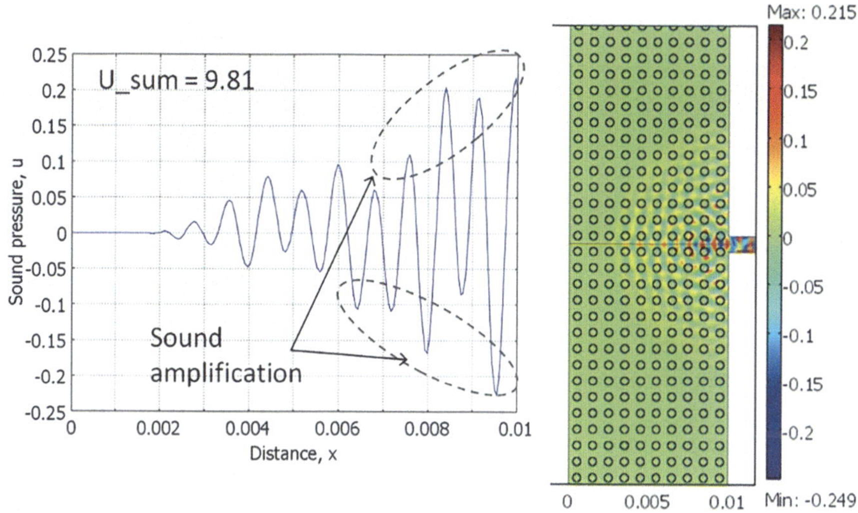

Fig. 2.54 On the RHS: the acoustic field in the layer of water with the uniformly distributed particles. On the LHS: the cross-sectional plot of the sound pressure along the longitudinal axis.

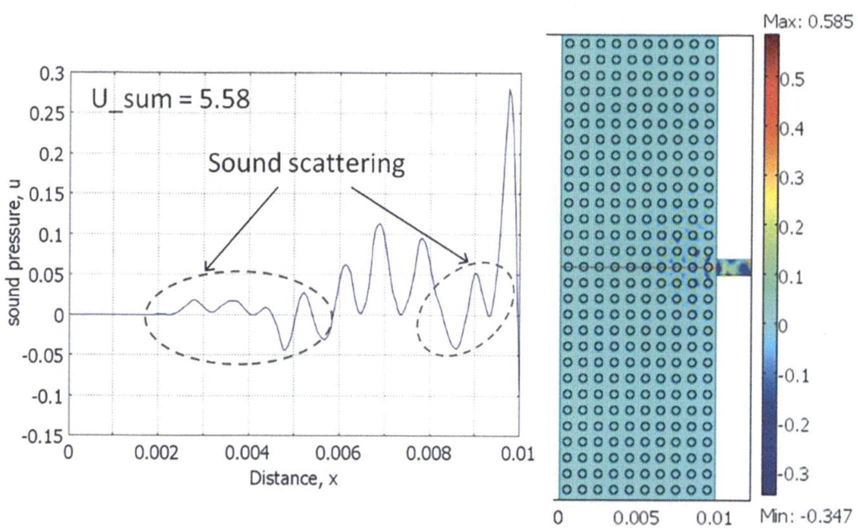

Fig. 2.55 On the RHS: the acoustic field in the layer of water with the uniformly distributed particles "shifted" upward, "blocking" the incoming sound beam. On the LHS: the cross-sectional plot of the sound pressure along the longitudinal axis.

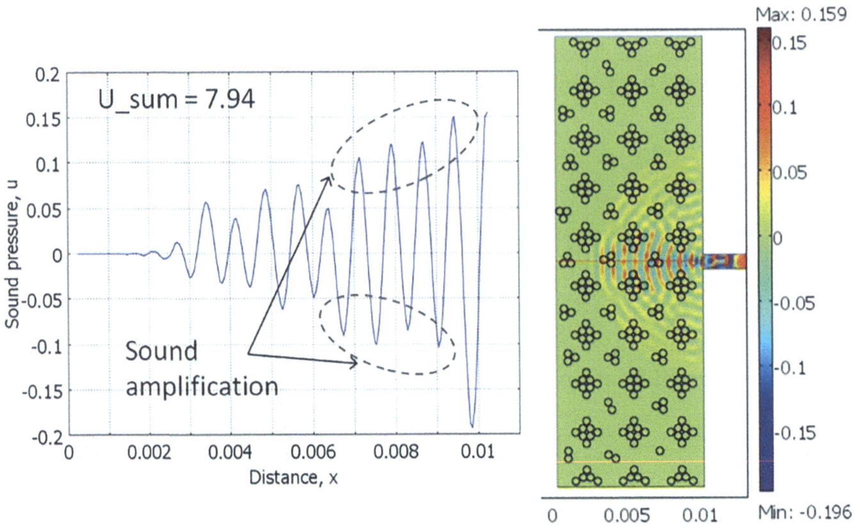

Fig. 2.56 On the RHS: the acoustic field in the layer of water with the particle clusters. On the LHS: the cross-sectional plot of the sound pressure along the longitudinal axis.

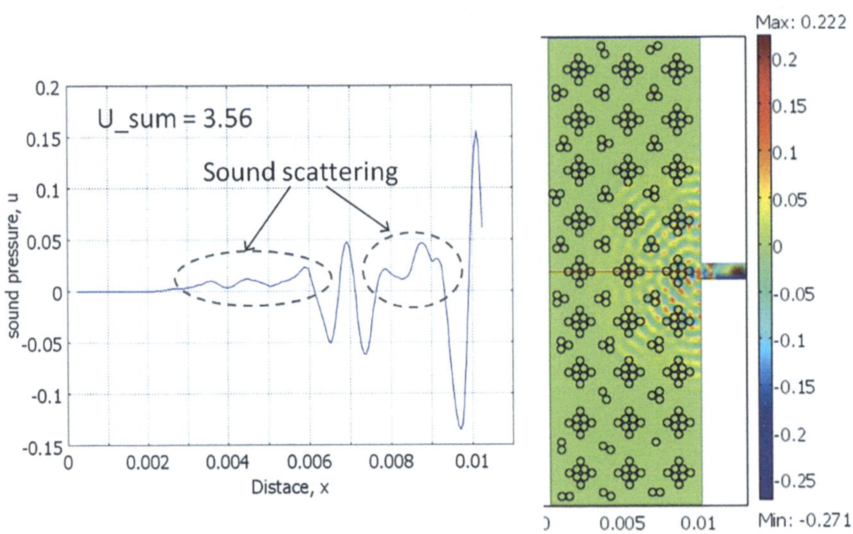

Fig. 2.57 On the RHS: the acoustic field in the layer of water with the particle clusters "shifted" upward, "blocking" the incoming sound beam. On the LHS: the cross-sectional plot of the sound pressure along the longitudinal axis.

Table 2.12 The calculated acoustic sound pressure values for fluid systems.

Simulation scenario	Absolute sound pressure (U_{sum})
1. Water, Fig. 2.53	6.96
2. Water with dispersed particles creating a "tunnel" for the ultrasonic wave to pass through, Fig. 2.54	9.81
3. Water with dispersed particles "blocking" the ultrasonic wave to pass through, Fig. 2.55	5.58
4. Water with clustered particles creating a "tunnel" for the ultrasonic wave to pass through, Fig. 2.56	7.94
5. Water with clustered particles "blocking" the ultrasonic wave to pass through, Fig. 2.57	3.56

2.6.5 Conclusions

- In order to understand how ultrasound waves propagate in water and water-particle systems, the time-dependent simulations were performed in COMSOL Multiphysics.
- The intrinsic, scattering, and structural sound losses composing the total sound attenuation were calculated for the particle suspensions and only the intrinsic sound loss for water, using the finite element software.
- The pattern of the simulated sound propagation in water is regular, while the particle presence introduced irregularities. This means that in addition to the sound absorption due to the particles, the sound scattering causing the sound pressure amplification or reduction is present, which depends on the spatial particle distribution.
- The particles that "block" the incoming sound wave, scatter it effectively (Cases 3 and 5 in Table 2.12), while the particles that create a "tunnel" for the wave to reflect, enhance its energy through the resonance effect (Cases 2 and 4 in Table 2.12). While the sound attenuation due to the absorption and scattering from particles is expected in real experiments, the sound amplification is a phenomenon, which is hypothetically possible for a static, "frozen-like" suspension, with a fixed particle arrangement that is favorable to amplify the sound.
- The structural sound losses are stronger for the particles joined in clusters (Cases 4 and 5, Table 2.12), than for the non-clustered dispersed particles (Cases 2 and 3, Table 2.12). However, the simulation result (Case 4, Table 2.12) showed that it is theoretically possible that the wave pressure can be increased by the clustering effect, compared to pure water.

- The simulations showed that the particles in water may either damp or amplify the sound to a higher degree than water. The first effect is more realistic, acting on the natural, chaotic mixing and arrangement of particles, damping the sound traveling through. Comparing the simulation results to the obtained experimental results, the sound loss due to particle presence is more realistic than the sound amplification. However, the sound amplification due to initiation of particle clustering is also possible (as in the experimental results, Fig. 2.43), which corresponds to Case 4.
- This simplified simulation approach does not consider real flow conditions, where often interactions between particles occur, as well as turbulent flow structures that are created at high flow speeds. However, it illustrates well the attenuation of sound waves in homogenous fluid and two-phase dispersions.

Competing Interests The author has no conflicts of interest to declare that are relevant to the content of this chapter. Part of the source material was reprinted from: (1) Applied Acoustics, 73 (2), B.M. Wrobel, R.W. Time, "Ultrasonic measurement and characterization of a low concentration system of solid particles in liquid, in high shear flow", pp. 117-131, February 2012, (2) Applied Acoustics, 72 (6), B.M. Wrobel, R.W. Time, "Improved pulsed broadband ultrasonic spectroscopy for analysis of liquid-particle flow", pp. 324-335, May 2011, with permission from Elsevier.

References

M.A. Ainslie, J.G. McColm, A simplified formula for viscous and chemical absorption in sea water. J. Acoust. Soc. Am. **103**(3), 1671–1672 (1998)

W.S. Ament, Sound propagation in gross mixtures. J. Acoust. Soc. Am. **25**, 638 (1953)

X. Bai, D. Wang, L. Zhen, M. Cui, J. Liu, N. Zhao, C. Lee, B. Yang, Design and micromanufacturing of focused piezoelectric ultrasound transducers for biological applications. Int. J. Extreme Manuf. **6**(6), 062001 (2024)

J.A. Bamberger, M.S. Greenwood, Measuring fluid and slurry density and solids concentration non-invasively. Ultrasonics **42**, 563–567 (2004)

J. Beckmann, J. Döring, J. Bartusch, U. Zscherpel, U. Ewert, A. Erhard, Time domain spectroscopy techniques for non-destructive contact less flaw detection and imaging in advanced materials, in *ECNDT. Poster 190*, (2006)

M.C. Bhardwaj, High transduction piezoelectric transducers and introduction of non-contact analysis. NDT.net, 5, 01 (The Largest Open Access Portal of Nondestructive Testing (NDT)) (2000). www.ndt.net/article/v05n01/bhardwaj/bhardwaj.htm

B.H. Birkofer, *Ultrasonic In-Line Characterization of Suspensions*. Diss. ETH No. 17331, Zurich, Switzerland, 2007

C.W. Brennan, *R2Sonic LLC Multibeam Training—Basic Acoustic Theory* (R2Sonic LLC, 2017). https://www.r2sonic.com/wp-content/uploads/2020/03/Basic-Acoustic-Theory.pdf

Bristol General Hospital, Ultrasonic imaging of the human body. Physics **62**, 671–722 (1999)

M.R. Davidson, D.J.E. Harvie, P. Liovic, Electrokinetic flow resistance in pressure-driven liquid flow through a slit-like microfluidic contraction, in *16th Australasian Fluid Conference, Australia*, (2007)

A.S. Dukhin, P.J. Goetz, Acoustic spectroscopy for concentrated polydisperse colloids with high density contrast. Langmuir **12**, 4987–4997 (1996)

A.S. Dukhin, P.J. Goetz, *Ultrasound for Characterizing Colloids. Particle Sizing, Zeta Potential, Rheology*, 1st edn. (Elsevier, New York, 2002)

A.S. Dukhin, P.J. Goetz, Bulk viscosity and compressibility measurement using acoustic spectroscopy. J. Chem. Phys. **130**(12), 1245191–1245193 (2009)

A.S. Dukhin, P.J. Goetz, *Characterization of Liquids, Dispersions, Emulsions, and Porous Materials Using Ultrasound*, 3rd edn. (Elsevier, Amsterdam, 2017)

A.S. Dukhin, P.J. Goetz, B. Travers, Use of ultrasound for characterizing dairy products. J. Dairy Sci. **88**, 1320–1334 (2005)

Endress+Hauser, *Flow Handbook. A Practical Guide: Measurement Technologies—Applications—Solutions* (Endress+Hauser Flowtec AG, Reinach, Basel, 2006)

EVIDENT, *High-Frequency Transducers. Thickness and Flaw Inspection Solutions* (2005). https://ims.evidentscientific.com/en/probes/single-and-dual-element/high-frequency-transducers

R.L. Gibson, M.N. Toksöz, Viscous attenuation of acoustic waves in suspensions. J. Acoust. Soc. Am. **85**(5), 1925–1934 (1989)

A.H. Harker, J.A.G. Temple, Velocity and attenuation of ultrasound in suspensions of particles in fluids. J. Appl. Phys. **21**, 1576–1588 (1988)

S. Haugen, S. Hodgson, Clamp on ultrasonic instruments in subsea applications. OTC **7746**, 471–477 (1995)

M.J. Holmes, N.G. Parker, M.J. Povey, Temperature dependence of bulk viscosity in water using acoustic spectroscopy. J. Phys. Conf. Ser. **269**, 012011 (2011). https://dx.doi.org/10.1088/1742-6596/269/1/012011

R.J. Hunter, *Introduction to Modern Colloid Science*, 1st edn. (Oxford University Press, New York, 1993)

International Organization for Standardization, *Measurement and Characterization of Particles by Acoustic Methods—Part 1: Concepts and Procedures in Ultrasonic Attenuation Spectroscopy*, 1st edn. (ISO 20998-1:2006(E), 2006)

T. Kobayashi, H. Tai, S. Kato, Measurement method of particle concentration and acoustic properties in suspension using a focused ultrasonic impulse radiated from a plano-concave transducer. Ultrasonics **44**, e491–e496 (2006)

M. Levy, H.E. Bass, R. Stern, *Modern Acoustical Techniques for the Measurement of Mechanical Properties* (Academic Press, San Diego, 2001)

R.L. Maharidge, *A Particle Sizing Technique Using Rayleigh Backscattered Ultrasonic Waves* (Sahio Petroleum Co., SPE 13643, 1985)

M. Manninen, V. Taivassalo, *On the Mixture Model for Multiphase Flow* (VTT Publications, Espoo, 1996)

F. Mbailassem, E. Gourdon, Q. Leclere, E. Redon, T. Cambonie, Sound absorption prediction of linear damped acoustic resonators using a lightweight hybrid model. Appl. Acoust. **150**, 14–26 (2019)

D.J. McClements, Ultrasonic characterization of emulsions and suspensions. Adv. Colloid Interf. Sci. **37**, 33–72 (1991)

P.A. Meyer, J.L. Rose, Guided wave inspection. NDT Database J. **2**(3) (1997). www.ndt.net/article/meyer2/meyer2.htm

NDT Products, Ultrasonic Transducers Explained: Your Ultimate Guide for Technology (2024). https://www.ndtproducts.ca/ultrasonic-transducers/

P. Nielsen, *Coastal Bottom Boundary Layers and Sediment Transport*, Advanced Series on Ocean Engineering, vol 4 (World Scientific, Singapore, 1992)

Y. Ozaki, T. Kawaguchi, Y. Takeda, K. Hishida, M. Maeda, High time resolution ultrasonic velocity profiler. Exp. Thermal Fluid Sci. **26**, 253–258 (2002)

H.A. Rabenjafimanantsoa, *Particle Transport and Dynamics in Turbulent Newtonian and Non-Newtonian Fluids*. PhD thesis, University of Stavanger, Norway, 2007

S.S. Rao, *Mechanical Vibrations*, 3rd edn. (Addison-Wesley Publishing Company, 1995)

W.T. Richards, The heating of liquids by the absorption of sound and its relation to the energy of intense high-frequency sound waves. Physics **17**, 611–616 (1931). https://www.pnas.org/doi/pdf/10.1073/pnas.17.11.611

V.J. Stakutis, R.W. Morse, M. Dill, R.T. Beyer, Attenuation of ultrasound in aqueous suspensions. J. Acoust. Soc. Am. **27**, 539–546 (1995)

G.G. Stokes, On the theories of the internal friction in fluids in motion, and of the equilibrium and motion of elastic solids. Trans. Camb. Philos. Soc. **8**(22), 287–342 (1845). https://archive.org/details/cbarchive_39179_onthetheoriesoftheinternalfric1849/page/n15/mode/2up

S. Temkin, Sound propagation in dilute suspensions of rigid particles. J. Acoust. Soc. Am. **103**(2), 838–849 (1998)

R.W. Time, *Analysis of Space and Time Structures in Two-Phase Flow Using Capacitance Sensors*. Diss. 7/1993, University of Bergen, Norway, 1993

R.J. Urick, A sound velocity method for determining the compressibility of finely divided substances. J. Appl. Phys. **18**, 983 (1947)

R. Winkler, *Sounding Out Flow. How Ultrasonic Flowmeters Work* (Omega Engineering, Inc., 2010). https://www.omega.co.uk/techref/pdf/SoundingOutFlow.pdf

B.M. Wrobel, R.W. Time, Improved pulsed broadband ultrasonic spectroscopy for analysis of liquid-particle flow. Appl. Acoust. **72**(6), 324–335 (2011). https://dx.doi.org/10.1016/j.apacoust.2010.11.013

B.M. Wrobel, R.W. Time, Ultrasonic measurement and characterization of a low concentration system of solid particles in liquid, in high shear flow. Appl. Acoust. **73**(2), 117–131 (2012). https://dx.doi.org/10.1016/j.apacoust.2011.07.006

A. Zaidni, S. Benjelloun, On sound dispersion in simple and multi-fluids due to thermal and viscous effects. Proc. Meetings Acoust. **45**(1) (2021). https://dx.doi.org/10.1121/2.0001554

Chapter 3
Complementary Research

3.1 Subproject Objectives and Challenges

3.1.1 The Subproject Objectives

- Designing and building a small-scale, rotational flow cell, which has an annular flow channel, for studying liquid-particle flow (the test cell's photos are in Fig. 2.4, Sect. 2.2.2).
- Choosing a suitable ultrasonic method for the measurement of the average flow speed of liquid-particle flow in the test cell.
- Verifying the ultrasonic results by the particle image velocimetry (PIV).
- Studying acoustic streaming in liquids.

3.1.2 The Subproject Challenges

- The rotational flow cell production by Forus Industri AS (Crisma Plastics) in Stavanger, Norway.
- Flow testing using the test cell.
- The flow velocity measurement of water-glass particle suspensions in the rotational test cell, using the ultrasonic transit-time difference (TTD) and particle image velocimetry (PIV) methods.
- The flow velocity measurement of water-glass particle suspensions in the small square-shaped cell (Fig. 2.3, Sect. 2.2.2), using the cross-correlation of ultrasonic backscattered signals.
- The determination of liquid-particle velocity profiles by the TTD method, supported by the PIV method.

B. Omylska, *Ultrasonic Measurement and Characterization of Liquid-Particle Flow*, https://doi.org/10.1007/978-3-032-03957-6_3

3.2 The Small-Scale Rotational Flow Cell

3.2.1 Applications

The small-scale rotational flow cell (Fig. 2.4 in Sect. 2.2.2, and Figs. 3.1 and 3.2) offers studying of the following applications:

- The axial (vertical) and transverse (horizontal) particle concentration profiles, along the flow channel's height and across the flow channel, respectively. The latter one is more challenging. Utilization of ultrasonic methods supported by the PIV method is recommended. The energy of an ultrasound pulse higher than 100 μJ is required for testing, due to the pulse traveling for a relatively long distance (through the box and flow channel parts of the cell), to obtain a strong enough signal for measurement.
- The axial and transverse velocity profiles of liquids and liquid-solid particle suspensions.
- Studying particle bed structures, using suitable ultrasonic methods and the PIV measurement method.
- Studying the distribution of dunes for various rotor speeds.
- Studying liquid-particle flow regimes in the cell.

3.2.2 Mechanical Design

The mechanics of the small-scale rotational flow cell was designed according to the front, top, and cross-sectional diagrams, Figs. 3.3 and 3.4. Materials used for building the cell were: transparent acrylic plastic for two cylinders, the cover and the box, and metal for the motor's shaft. The measurement volume of the cell is $V = 2860$ mL, the height is $H = 86$ mm, and the width is $W = 45$ mm.

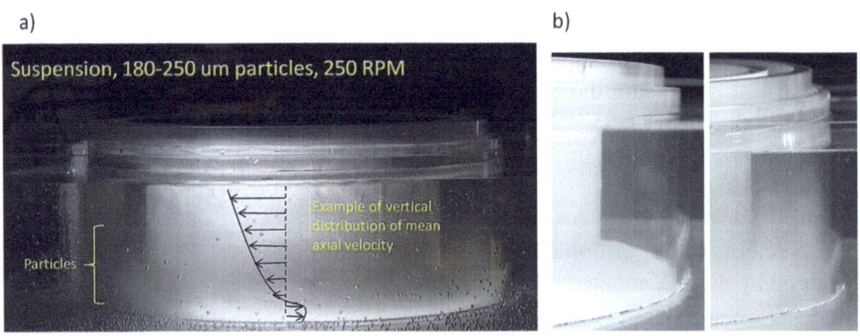

Fig. 3.1 (**a**) The suspension of the glass particles (180–250 μm, $35.5\%_{wt} \approx 3.97\%_{vol}$) in water, for the rotor speed of 250 RPM (3.14 m/s). (**b**) The cross-sectional views at static and flow conditions.

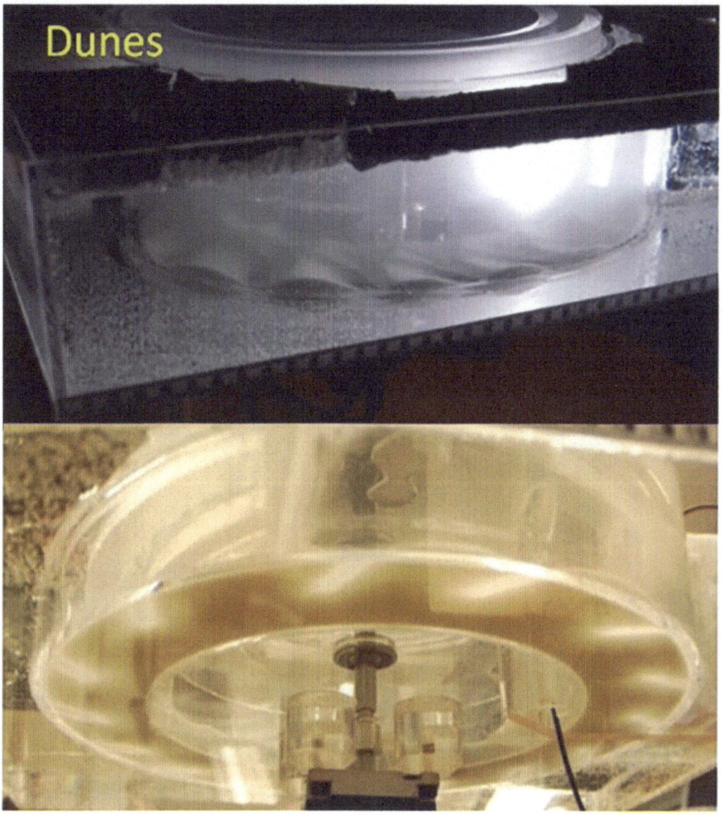

Fig. 3.2 The glass-particle dunes along the circular flow channel, in the small-scale rotational flow cell.

3.2.3 Motor Mechanical Design

3.2.3.1 The Driving System of the Test Cell

The driving system of the test cell consisted of the following components:

- The MIS231 QuickStep stepper motor from Farnell, delivered by JVL Industri Elektronik A/S, with the voltage of 12 V, the current of 0.6 A, and the holding torque of 550 mN·m.
- The driver for the stepper motor (30 V DC, 2 A) from Farnell. The current and voltage of the driver must exceed the current and voltage of the motor.
- The backplane card from Farnell with the connectors for a driver and a power supply.

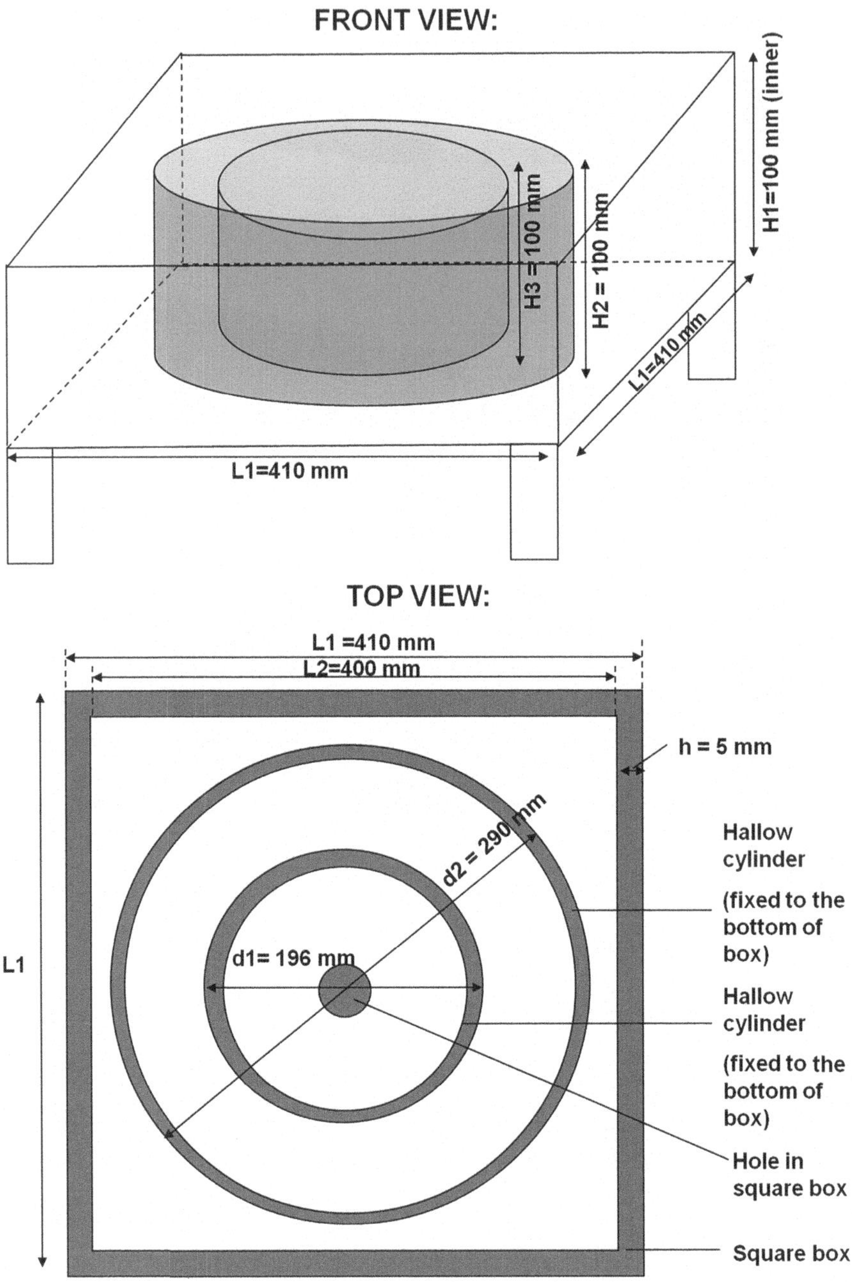

Fig. 3.3 The front and top views of the small-scale rotational flow cell design.

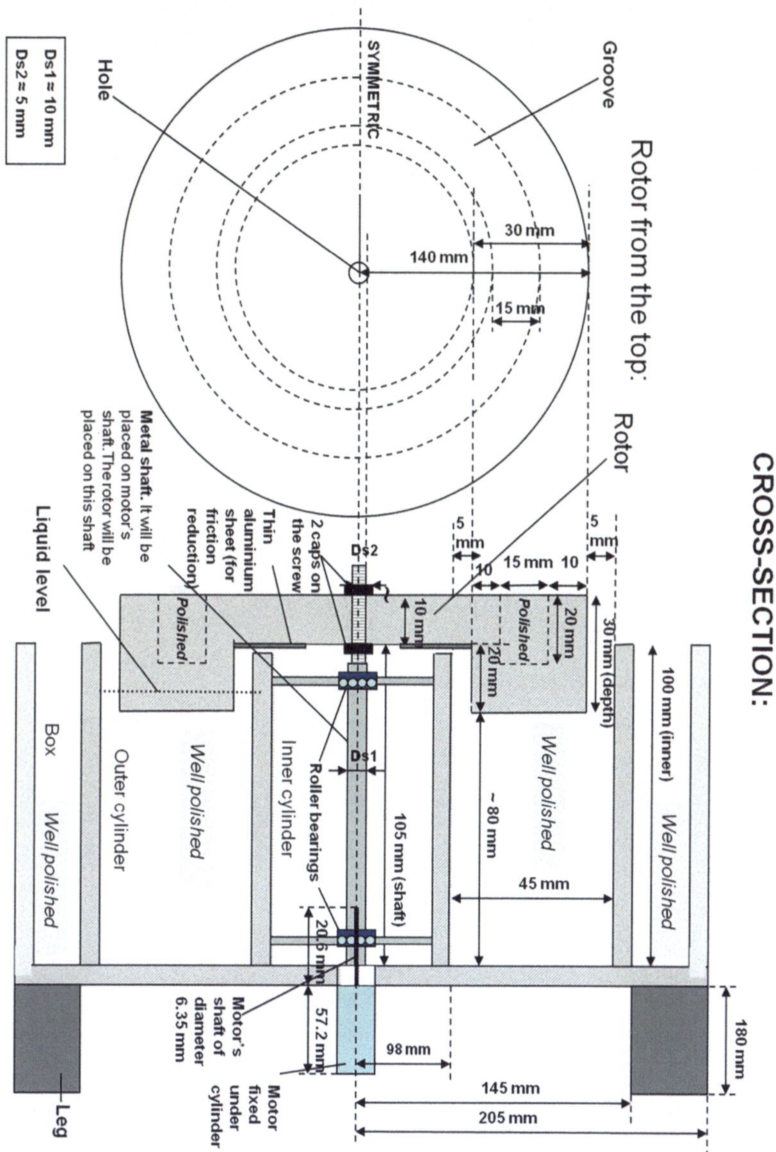

Fig. 3.4 The cross-sectional view of the small-scale rotational flow cell design.

3.2.3.2 The Test Cell's Motor Mechanical Design

The choice of the stepper motor was based on the design calculations presented in Tables 3.1, 3.2, 3.3, 3.4, 3.5, 3.6, and Fig. 3.5. The following parameters were calculated: the required linear rotor frequency, the displacement of the rotor by one pulse, the maximum total torque consisting of the viscous and inertia torques (Fig. 3.6).

The motor needs to overcome the torque, which consists of the viscous part related to fluid viscosity and the inertia part of the rotating plate, Fig. 3.6. The viscous torque is insignificantly small, compared to the inertia torque. The calculated maximum torque is 660 mN·m, for the rotor velocity of 1 m/s, however, this is the torque to overcome only at the start-up. In the later phase, the torque will be consequently smaller due to the rotor's inertia. It was decided that the motor holding torque of 550 mN·m is suitable for the test cell.

Table 3.1 The calculated angular and linear rotor frequencies.

Average radius R_{av} (m)	Motor rotational speed U_{RPM} (rev/min)	Rotor speed[a] U (m/s)	Angular rotor frequency[a] ω (rad/s)	Linear rotor frequency[a] f_0 (Hz)
0.125	40	0.5	4	0.64
0.125	80	1	8	1.27

[a] The related equations are: $U = 2\pi R \cdot U_{RPM}$, $\omega = U/R$, $f_0 = \omega/(2\pi)$

Table 3.2 The calculated required rotor frequency.

Linear rotor frequency f_0 (Hz)	Rotor frequency steps $f_{rot\text{-}steps}$ (no. of steps/revolution or no. of pulses/rotation)	Required rotor frequency[a] $f_{rot\text{-}req}$ (Hz)
0.64	200	127.32
1.27	200	254

[a] The related equation is: $f_{rot\text{-}req} = f_0 \cdot f_{rot\text{-}steps}$

Table 3.3 The calculated friction force and viscous torque.

Fluid viscosity μ (Pa·s)	Radius R_1 (m)	Radius R_2 (m)	Area[a] A (m^2)	Height H (m)	Speed U (m)	Friction force[a] F_w (N)	Viscous torque[a] T_v (mN·m)
0.001 (water)	0.105	0.145	0.03	0.1	0.5	1.57e-4	0.02
0.001	0.105	0.145	0.03	0.1	1	3.14e-4	0.04
0.01 (a slurry)	0.105	0.145	0.03	0.1	0.5	1.57e-3	0.2
0.01	0.105	0.145	0.03	0.1	1	3.14e-3	0.4

[a] The related equations are: $A = \pi\left(R_2^2 - R_1^2\right)$, $F_w = \mu U(A/H)$, $T_v = F_w(R_1 + R_2)$

Table 3.4 The calculated displacement of the rotor plate by a single pulse.

Rotor frequency steps $f_{rot\text{-}steps}$ (no. of pulses/rotation)	Rotation[a] p (rotation/pulse)	Circumference[a] C (m)	Displacement[a] x (mm)
200	0.005	0.91	4.4

[a] The related equations are: $p = 1/f_{rot\text{-}steps}$, $C = 2\pi R_2$, $x = pC$

Table 3.5 The calculated torque due to the mass of the rotor plate (it is related to its inertia).

Plate volume V (m^3)	Acrylic density ρ (kg/m^3)	Plate mass m (kg)	Rotor speed U (m)	Angular speed ω (rad/s)	Inertia moment[a] I ($kg{\cdot}m^2$)	Inertia torque[a] T_i (mN·m)
0.00164	1200	1.96	0.5	4	0.021	165
0.00164	1200	1.96	1	8	0.021	660

[a] The related equations are: $I = 0.5mR_2^2$, $T_i = 0.5I\omega^2$

Table 3.6 The calculated maximum total torque.

Rotor speed U (m/s)	Fluid viscosity μ (Pa·s)	Maximum total torque $T_v + T_i$ (mN·m)
0.5	0.001 (water)	0.02 + 165
1	0.001	0.04 + 660
0.5	0.01 (a slurry)	0.2 + 165
1	0.01	0.4 + 660

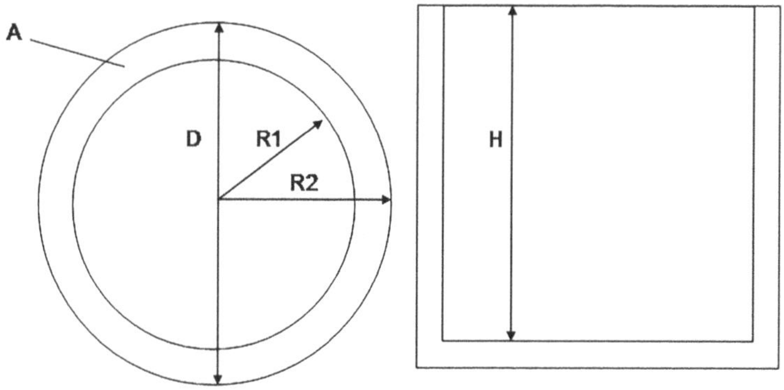

Fig. 3.5 The design parameters of the rotating cover and the flow channel of the small-scale rotational flow cell.

3.2.4 Ultrasonic Transducer Configurations

There are two alternatives for placing ultrasonic transducers for measurements (Fig. 2.4, Sect. 2.2.2). The first configuration is for the transducers placed externally on the box's walls of the test cell, which enables the measurement across the annular channel and the space created by the box and cylinder, which is filled with water.

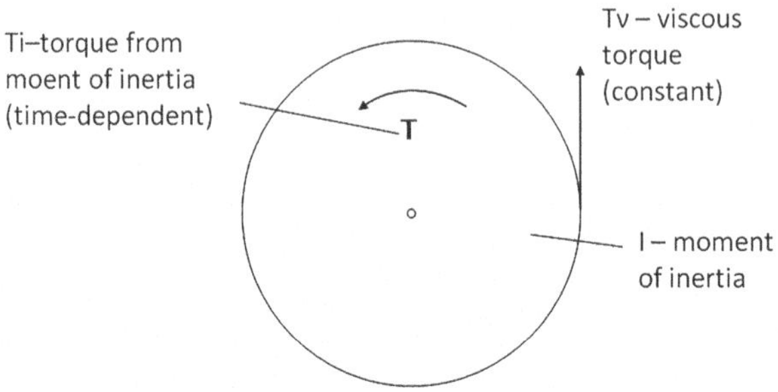

Fig. 3.6 A sketch showing the inertia and viscous torques of the rotating cover, driving flow from the "top," in the small-scale rotational flow cell.

The second configuration is for two transducers that adhere to the external cylinder's wall, and the measurement is only within the flow channel (the transmitted sound beam will be reflected from the inner cylinder, back to the receiving transducer). Based on the initial testing using the transit-time difference (TTD) technique, it turned out that the first configuration gives reliable results. A downside of the second configuration is that the sound beam's path is too short to be influenced by flow in the test cell. This is not the case for the first alternative, due to a long enough distance for the sound to travel in the "with the flow" and "against the flow" directions.

3.2.5 *Briefly on the Rotational Flow Cell Testing*

Several adjustments related to the proper operation of the rotational flow cell were implemented, after the cell was initially tested. The cell speed is remotely controlled by the step motor software, MacTalk. The maximum, limiting speed is 270 RPM (3.4 m/s), due to the appearance of a large air ring right under the rotating cover, when running the cell at the higher speed, causing an unstable flow pattern. The suitable running current of 0.5–0.8 A and the voltage of 5 V, give the power of 2.5–4 W. These parameters protect the motor from overheating. The cover mounting was improved by putting metal discs, both on the shaft and under the cover, to minimize its vertical movement and smoothing the rotational movement. Additional parts were mounted to the cell, which are used for holding ultrasonic transducers at a chosen level during tests.

3.2.6 *The Simple Couette Flow Theory*

3.2.6.1 The Simple Couette Flow Theory

The equation for the angular velocity of a rotating object, U_{RPM} (rad/s or s^{-1} in the SI base units), is the following:

$$U_{\mathrm{RPM}} = \frac{U}{C} = \frac{U}{2\pi r} \tag{3.1}$$

where U is the linear velocity of a point in the rotor (m/s), C refers to the circumference of a circle that this measuring point performs around the axis of rotation (m).

For the simple Couette flow, the shear stress for Newtonian fluid, τ_{xy}, at the surface element parallel to a flat plate, at the point y (Pa), is given by:

$$\tau_{xy} = \mu \frac{\partial u}{\partial y} \tag{3.2}$$

where $\partial y = y_2 - y_1$ is the distance measured from the bottom of the plate (m), where y_1 is equal to 0 m for the rotational test cell, and $\partial u = u_2 - u_1$ is the corresponding velocity of fluid (m/s), where u_1 is equal to 0 m/s, and μ is the dynamic viscosity of fluid (Pa·s).

The fluid velocity, u (m/s), is calculated by Munson et al. (2006):

$$u = U \frac{y}{h} \tag{3.3}$$

There is no pressure drop along the annular flow channel due to a closed system; therefore, $\partial p/\partial x$ is negligible (Munson et al. 2006).

3.2.6.2 The Theoretical Flow Characteristics

The theoretical flow calculations are presented in Fig. 3.7, and conclusions are the following:

- There is constant fluid shear stress along the flow channel's height.
- The maximum flow velocity is at the flow channel's top and the minimum velocity (zero) at the flow channel's bottom.
- The mean linear flow velocity values are 0.01–3.4 m/s (1–270 RPM). They are measured at the flow channel's top, for the average radius of the flow channel. The flow velocities correspond to the Reynolds numbers of 450–175,500. They correspond to laminar and turbulent flow regimes.

3.2.7 Flow Characteristics

Based on the initial flow tests, visual observations and high-speed camera recordings, the flow characteristics in the flow cell are the following:

- Ideal seeding particles for the PIV measurement are the small diameter 20–40 μm glass beads, added to water in the amount not exceeding 10 g. Heavier particles

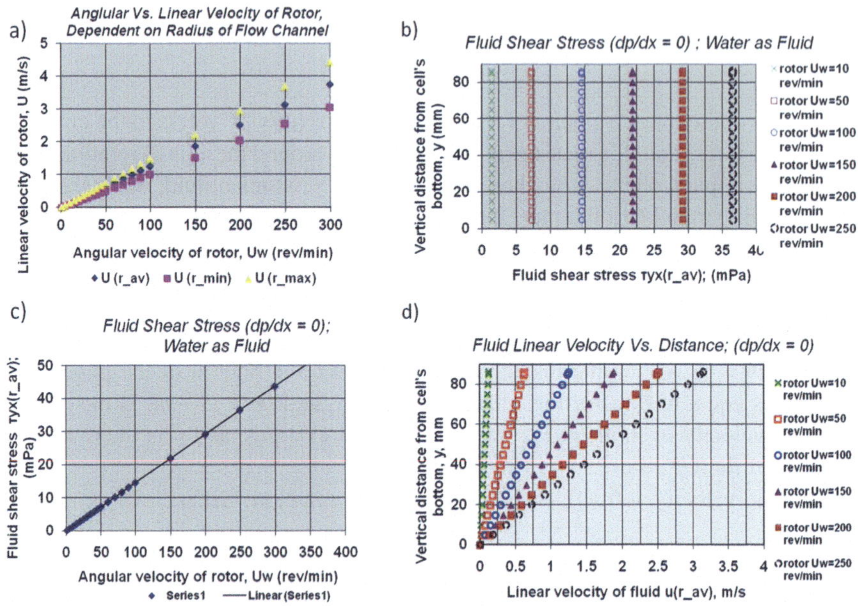

Fig. 3.7 Calculated parameters for the simple Couette flow in the small-scale rotational flow cell. (**a**) Shows the relationship between the rotor angular and linear velocities, for different radii of the flow channel. (**b**) Illustrates the fluid (water) shear stress versus the vertical distance from the cell's bottom, for various rotor speeds. (**c**) Depicts the rotor angular velocity against the shear stress for water. (**d**) Displays the fluid linear velocity versus the distance, for various rotor speeds.

will be swept toward the inner wall of the cylinder, and lighter particles will float in water, like neutrally buoyant particles.

- An ideal aqueous suspension is created for the rotor speed of 250 RPM (3.14 m/s) with the 180–250 μm, 2.64 g/cm^3 glass particles, in the amount of 300 g (corresponding to the particle concentration of 35.5%$_{wt}$ or 3.97%$_{vol}$).
- A gas bubble ring appears just under the rotor. For the flow speed values higher than 250 RPM, the ring becomes more developed and unstable.
- In the case of a flowing suspension, in addition to the main flow along the longitudinal direction, there is also the secondary transverse flow at the cell's bottom, resulting in particles accumulated at the inner wall of the flow channel. The secondary flow type has a disturbing effect on the main flow, which can be seen only when heavy particles are placed in the cell, and it cannot be seen in the case of pure water flow. Figure 3.8 shows how the transverse flow pattern can look in the rotational flow cell. Another possible side flow pattern would take a toroidal or a helix shape, which would be pronounced in the walls' area and not in the main flow, Fig. 3.9.
- Creeping dunes are formed at the low rotor speeds of 30–100 RPM (0.37–1.25 m/s), for high amounts of the 106 and 180–250 μm sized glass particles. They create a thick bed (higher than 1 cm) on the cell's bottom. After flow is established, the dunes tend to be symmetrically distributed along the flow channel.

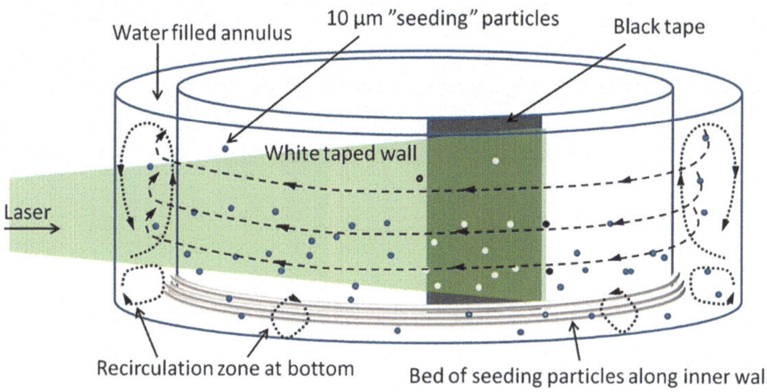

Fig. 3.8 Flow patterns in the small-scale rotational flow cell, for the cell filled with water and light "seeding" particles.

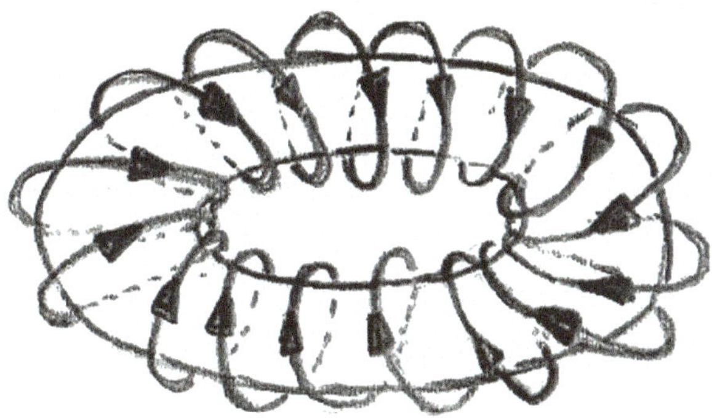

Fig. 3.9 A hand-drawn sketch of a toroidal coil with multiple loops. The coil is depicted with arrows indicating the direction of a flowing liquid-particle suspension, which loops around the torus representing a flow channel. The sketch illustrates the flow pattern, commonly present in the fluid dynamics studies of rotational systems.

3.2.8 Conclusions

- The small-scale rotational flow cell offers a wide range of possibilities for measurement and characterization of liquid-particle flow, for various flow speeds (and flow regimes), such as suspensions and dunes. The flow of a water-glass particle suspension consists of the main and secondary flow patterns in the cell.
- The concentration and velocity profiles of liquid-particle systems can be measured in the flow cell. Acoustic and laser-based methods can be used for testing and measurement, supported by data science programs for digital signal post-processing. For example, the particle image velocimetry (PIV) method can be used for the determination of a particle concentration profile, complemented by images acquired by a high-speed camera.
- The theoretical characteristics of the simple Couette flow for Newtonian fluid, such as water, were calculated for the test cell's dimensions. The constant shear stress values for fluid, along the flow channel's height, vary with the speeds of

the rotor. There is a linear velocity distribution for fluid in the flow cell, with the maximum value at the cell's top and the minimum value (zero) at the cell's bottom.

- The rotational small-scale test cell was successfully designed and produced with the support of multiple stakeholders from Norway and France. The mechanical design of the rotational flow cell included the calculations for choosing a suitable stepper motor.

3.3 The Transit-Time Difference Measurement of Flow Velocity Profiles

3.3.1 Subproject Objectives and Challenges

3.3.1.1 The Subproject Objectives

The objective of this study is to perform the feasibility measurement of the mean axial flow velocity profiles for liquid and liquid-particle flows, in the small-scale rotational cell (Sect. 3.2), using the ultrasonic transit-time difference (TTD) method. The flow velocity measured by the TTD method could be compared with the flow velocity measured by the particle image velocimetry (PIV) method.

3.3.1.2 The Subproject Challenges

- Obtaining readable ultrasonic signals across the annular flow channel of the small-scale rotational flow cell, for water and water-glass particle flowing suspensions, as well as for static reference conditions for the cell filled with water.
- The ultrasonic measurement of the average flow speed using the TTD method, for ultrasound waves traveling "with the flow" and "against the flow".
- Recording the videos and images of flowing water and water-glass particle suspensions, using a laser and a high-speed video. They will be used for the determination of the flow velocity profiles using the PIV technique.

3.3.2 Flow Tests Using the Transit-Time Difference

Tests were performed for water and the glass particle suspension (180–250 μm, $35.5\%_{wt}$), in the annular flow channel of the small-scale rotational test cell. The purpose of the tests was the determination of the mean axial flow velocity distributions, using the ultrasonic transit-time method. Two independent tests were carried out, in order to validate the results. The tests were performed for the configuration of transducers presented in Fig. 2.4 (the first alternative) in Sect. 2.2.2, where the transducers are placed on the external square part of the test cell. In this configuration, ultrasounds from the transmitting transducer pass through the system of multiple acrylic walls and static "standing" water, reaching the circular flow channel, then they travel through the second part of the "standing" water and are received by the opposite receiving transducer. The "standing" water is used for transmitting the pulse to the flow channel and for creating an ability to look through the cell, in the space between the outer cylinder and square box. During the tests, the transducers' positions were changed along the box's height, from 8 cm (at the top) to 1 cm (at the bottom) with the step of 0.25 cm, for the purpose of the measurement of velocity profiles. The speed of the rotor was set to 250 RPM (3.14 m/s). The theoretical, calculated mean axial flow speed values are 0 m/s at the flow channel's bottom, and 3.14 m/s at the flow channel's top, assuming that the flow speed is considered in the middle of the width of the channel, Fig. 3.7d. The shear stress is constant along the flow channel's height and is equal to 36.5 mPa, Fig. 3.7b. The spike pulses (500 Hz and 100 μJ) generated by the pulser-receiver were used in the through-transmission test. The signal was collected by the ultrasonic preamplifier (and not by the pulser-receiver) due to desired signal amplification. The absolute positions (times) of the ultrasonic pulses were recorded, where t_{WF} is the time of an ultrasonic pulse traveling "with the flow" direction (s) and t_{AF} is the time of an ultrasonic pulse traveling "against the flow" direction (s), Fig. 3.10. The difference between them was calculated by $DT = t_{AF} - t_{WF}$, Fig. 3.11a, c. The flow velocity, u, measured by the transit-time difference (TTD) method (m/s), for the sound beam traveling in the direction parallel to the flow direction, is calculated by the equation:

$$u = \frac{L}{2} \cdot \frac{\left(t_{AF} - t_{WF}\right)}{\left(t_{AF} \cdot t_{WF}\right)} \tag{3.4}$$

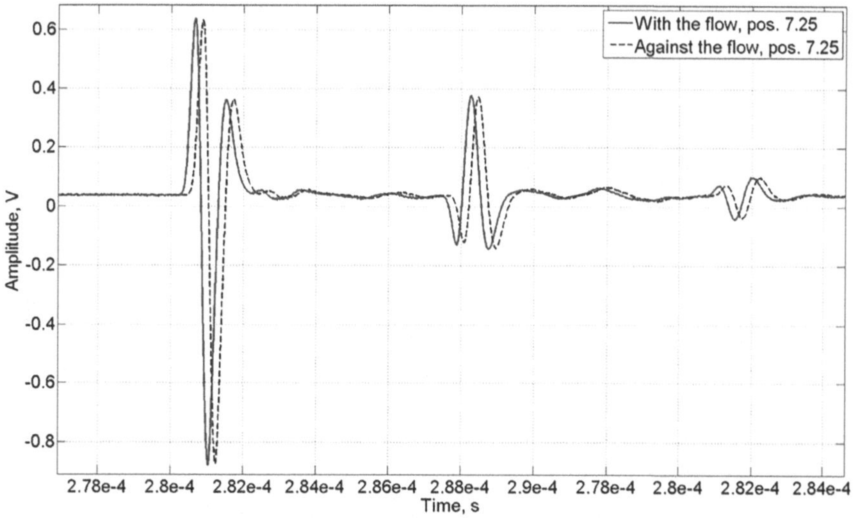

Fig. 3.10 The signals for flow speed measurement using the transit-time difference method.

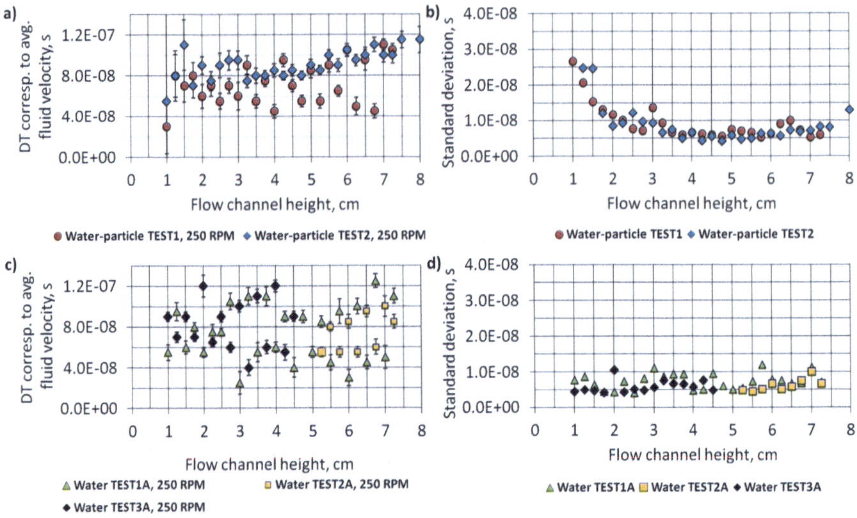

Fig. 3.11 The transit-time difference ultrasonic measurement of flow velocities for (**a**) the water-particle suspension (180–250 μm, 35.5$\%_{wt}$) and (**c**) water. The corresponding standard deviations are in (**b**) and (**d**).

where *L* is the distance between the receiving and transmitting transducers (m).

For the averaging purpose, 20 signal samples for the cases "with" and "against the flow" were recorded for each single measurement, and the corresponding standard deviation values were calculated, Fig. 3.11b, d.

The measured flow velocity values are verified using the particle image velocimetry in Sect. 3.3.6. It is important to note at this stage that the secondary flow in the transverse cross-section is present, in addition to the main flow, along the axial cross-section of the flow channel, Fig. 3.8.

Conclusions are the following:

- The flow speed values for the particle suspension measured by the transit-time difference method vary between 0.32 m/s (measured at 7 cm from the cell's bottom) and 0.1 m/s (measured at 1 cm from the cell's bottom), Fig. 3.12a. Similar values were obtained for pure water, however, the values do not clearly decrease toward the cell's bottom. The measured flow speed values are lower than theoretically calculated, due to the secondary flow slowing down the axial flow, and the gas ring under the rotor, reducing the transferred dragging force from the rotor to water.
- Generally, the time-of-flight difference values, *DT*, which correspond to the mean fluid axial velocity for the case of water-particle flow, slightly decrease toward the channel's bottom, Fig. 3.11a. This agrees with the calculated theoretical predictions, Fig. 3.7.

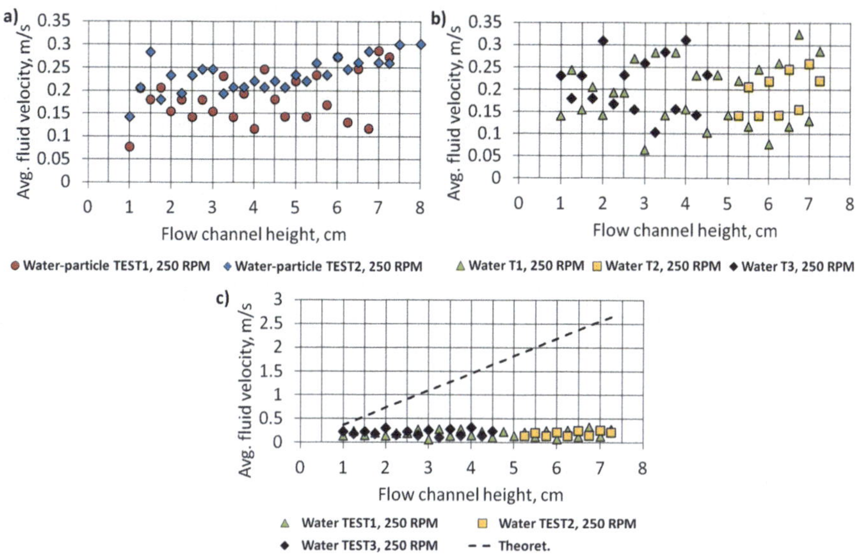

Fig. 3.12 The data from Fig. 3.11 converted to average fluid speed values (m/s), corresponding to the mean axial flow velocity distributions for: (**a**) the water-particle suspension (180–250 µm, 35.5%$_{wt}$) and (**b**) water. The comparison of the flow velocity data with the theoretical values (**c**).

- The standard deviation values for the case of water-particle flow were calculated based on 20 samples. The values increased at the channel's bottom, where the particle concentration is the highest. This is due to larger (and clearly visible during the tests) flow fluctuations at the bottom of the cell, Fig. 3.11b.
- The time-of-flight difference values, *DT*, for the case of water flow seem to decrease toward the channel's bottom, however, this is not evident, Fig. 3.11c. The large scattering of the data points indicates the large flow fluctuations along the whole channel's height. Normally, greater scattering would be expected for the flowing water-particle suspension, rather than for water, because of the particle inertia, whereas the opposite result was obtained. The associated hypothesis is that the distribution of the kinetic energy for the mixture is shifted toward the top of the flow channel. This is due to the particles occupying the space at the bottom of the flow channel. This means that the kinetic energy is damped by the particles at the bottom of the cell. This would not be the case for pure water, where the kinetic energy is more evenly distributed along the channel's height, which results in the significant scattering of the data points. This hypothesis is confirmed by the sensitivity test result for flow fluctuations in Sect. 3.3.3.
- The standard deviation values for the case of water flow are approximately constant over the channel's height, Fig. 3.11d. This confirms that high amount of particles caused the significant flow fluctuations at the channel's bottom.

3.3.3 *Ultrasonic Sensitivity to the Flow Fluctuations*

Fluctuations for water flow in the small-scale rotational flow cell were studied, by running the transit-time difference (TTD) measurements over a long time. Each of six series was taken every 1.5 min, which gives the total measurement time of 9 min. For comparison, each series was acquired during the interval of 30 s for the "with the flow" and "against the flow" directions. The measurements were performed at the fixed transducers' positions of 3.5 and 1.25 cm, measuring from the cell's bottom. The measurements were performed for both static and flow conditions for the rotor speed of 250 RPM (3.14 m/s).

The measured average time difference values, *DT*, between the pulses acquired "with the flow" and "against the flow" are presented in Fig. 3.13. For the transducers' position of 3.5 cm, the time difference is $7.08 \times 10^{-8} \pm 1 \times 10^{-8}$ s (corresponding to the velocity of 0.18 ± 0.027 m/s). For the transducers' position of 1.25 cm, the time difference is $7.58 \times 10^{-8} \pm 9.2 \times 10^{-9}$ s (corresponding to the velocity of 0.19 ± 0.023 m/s).

It can be assumed that the average flow fluctuation between each of six flow series is approximately 2×10^{-8} s (0.054 m/s), for the sound beam sent alongside the flow direction. This is a small number comparing to the test result for the flow of water, Fig. 3.11c, and it is much below its size of the standard deviation. This leads to the conclusion that the effect of data scattering along the channel's height, is indeed due to the turbulent flow structures, which occurred locally, and not due to the natural flow fluctuations Fig. 3.13. For comparison the static water tests were also performed, Fig. 3.13. The flow fluctuations study confirms very good sensitivity of the ultrasonic measurement system for flow measurement.

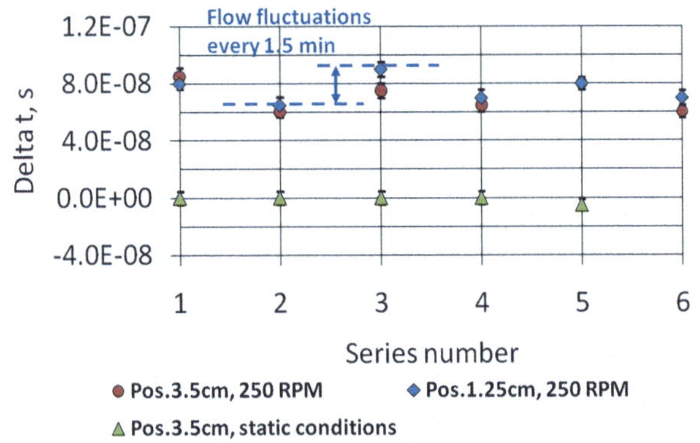

Fig. 3.13 The study of the ultrasonic sensitivity to flow fluctuations. The transit-time difference measurements were taken within the total time of 9 min, for the rotor speed of 3.14 m/s. Each test of the whole series was taken every 1.5 min.

3.3.4 Ultrasonic Sensitivity to the Rotor Speeds

An additional test using the transit-time difference (TTD) method was performed for the flow of water, for the rotor speeds from 25 to 275 RPM (0.30–3.45 m/s). The purpose was to verify the measurement sensitivity of the ultrasonic system to the rotor speeds.

The time-of-flight difference values, *DT*, for various rotor speeds are presented in Fig. 3.14. The results indicate on very good measurement sensitivity. The values increase proportionally to the rotor speeds, for the fixed transducers' positions of 5 and 6.5 cm (in the upper part of the flow channel). The minimum detected speed was 5 RPM (0.06 m/s). In the case of the transducers located in the lower part of the channel, there is a deviation from the straight line, above the rotor speed of 125 RPM, which is probably due to large flow fluctuations created by the local secondary flow structures. These flow fluctuations are due to the recirculation zone presence at the bottom of the cell, Fig. 3.8. The standard deviation values were calculated based on 20 signal samples. They are small, of the size of 5×10^{-9} s, and nearly constant for all the rotor speeds. These results can be compared with the results acquired for the case of water at static conditions, Fig. 3.13. The results point to good measurement sensitivity and accuracy of the ultrasonic system.

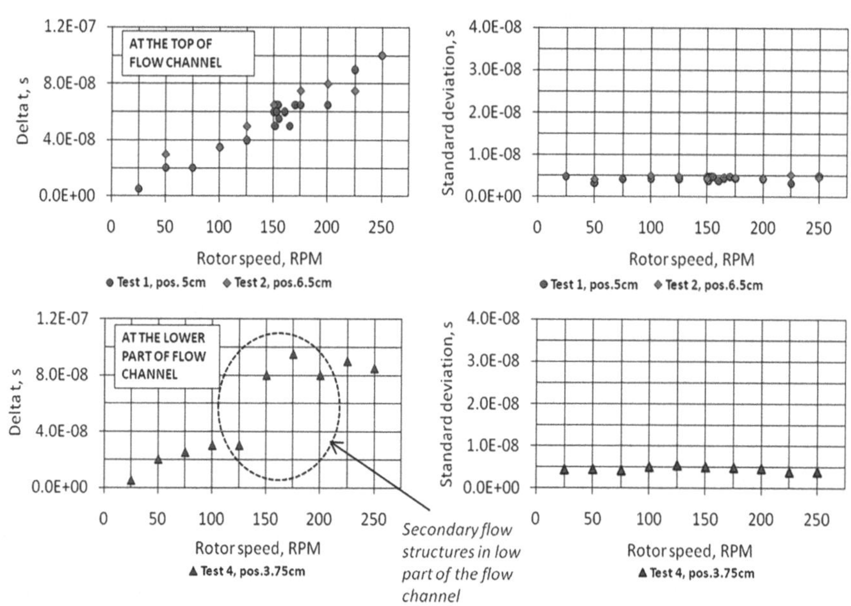

Fig. 3.14 The measured ultrasonic transit-time difference values for water flow, for the rotor speeds of 25–275 RPM (0.30–3.45 m/s), measured at the cell's bottom and top. The corresponding standard deviation values are on the RHS.

3.3.5 Sound Speed Dependent on Temperature

A change in temperature is important to consider, due to its impact on sound speed measurement. The temperature change of 0.1 °C changes the sound speed in water by 0.2 m/s (Fig. 8, Wrobel and Time 2012). Such change impacts the measurement reliability, when the ultrasound measurement is performed only in one direction.

The temperatures of the whole testing system (consisting of the motor, water inside the flow channel, static water inside the area between the square box and the flow channel), changed slightly during the whole duration of the tests lasting a few hours (with the average $\Delta T_{\mathrm{motor}} \approx 9$ °C, $\Delta T_{\mathrm{water-in}} \approx 0.4$ °C and $\Delta T_{\mathrm{water-out}} \approx 1$ °C). The temperature deviations influenced the absolute positions of the ultrasound signals, when comparing times at the beginning and at the end of the tests. However, it is important to underline that the temperature effect is neutralized, according to the transit-time difference measurement principle (for the simultaneous measurements in the "with" and "against the flow" directions).

In the tests, the continuous measurement of 20 signal samples was a standard procedure, which was performed separately for the sound traveling "with the flow" and "against the flow." In practice, changing the direction of the ultrasound beam's travel path was performed, by switching the connectors of the transmitting and receiving transducers. The time length of taking 20 signal samples was from 0.5 to 1 min. This is close to the simultaneous measurement, especially for the whole test lasting a few hours.

To verify, if temperature had any effect on the performed measurements, the ultrasound test for water at static conditions was run in the "with" and the "against the flow" directions. The test lasted for 95 min. and the measurements were performed at the cell's heights from 7.25 to 3.5 cm.

Results are presented in Fig. 3.15. The temperature of "standing" water outside the flow channel changed from 17.3 to 18.4 °C, which reduced the time of receiving the signal acquired "with the flow" direction by 1.44×10^{-6} s (the corresponding reduction in the sound speed is 2.2 m/s). However, it is important to underline that the time-of-flight difference values are equal to zero for all the test points (the

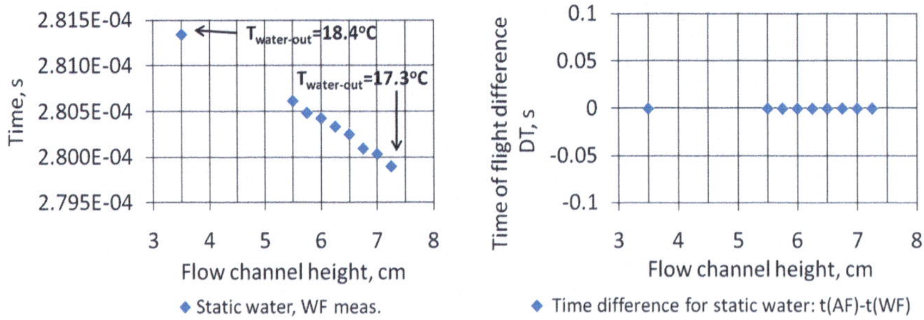

Fig. 3.15 On the LHS: the effect of temperature on the sound speed measurement for water. On the RHS: the time-of-flight difference values for the transit-time difference method.

subtraction of the absolute positions of the signals acquired "with the flow" and "against the flow" resulted in zero signal shifts), Fig. 3.15 on the right-hand side. This indicates no risk or a very low risk of the measurement error due to the temperature effect.

3.3.6 The Transit-Time Results Validation by the Particle Image Velocimetry

The verification of the transit-time difference ultrasonic measurement results (Sect. 3.3.2) was performed using the particle image velocimetry (PIV) method. For this purpose, images were recorded by a high-speed camera. A sheet of green laser light was illuminating the axial cross-section (along the flow channel's circumference, the mid-width), as well as the transverse cross-section (across the flow channel's width). Examples of the acquired images for a water-particle suspension, and water flow with seeding particles, are presented in Figs. 3.16 and 3.17. The 180–250 μm glass beads of the concentration of $35.5\%_{wt}$ were used for the suspension. The 20–40 μm glass beads added to water in the amount of 10 g, were used as the seeding particles.

The initial processing of the recorded images revealed that it is possible to cross-correlate them using the PIV method. The image processing was performed using the MatPIV 1.6.1 toolbox in MATLAB. In Fig. 3.18, an example of the calculated velocity vectors for the particle suspension is presented. The instantaneous, local velocity profile in Fig. 3.19 was determined, at the chosen position x of 440 px along the channel's height, based on two images acquired one after another, separated by the 2 ms time interval. The calculated velocity is related to movements of the glass particles. The estimated velocity values vary between 0.3 and 0.7 m/s in

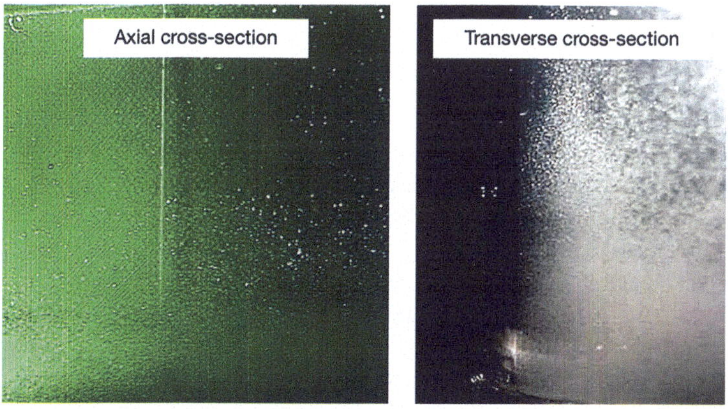

Fig. 3.16 The high-speed camera images for the flowing suspension of water with the glass particles (180–250 μm, $35.5\%_{wt}$).

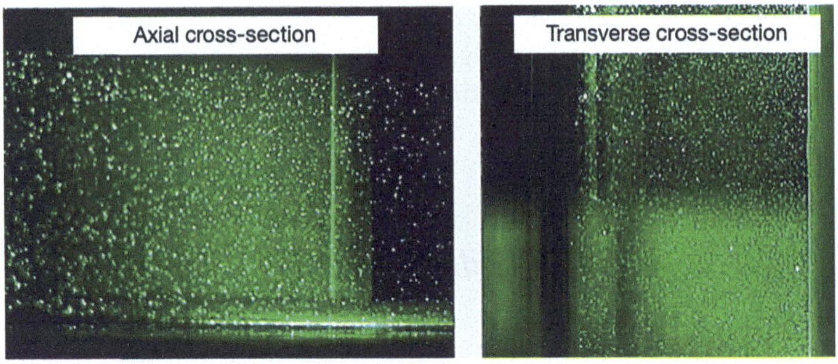

Fig. 3.17 The high-speed camera images for water with the 20–40 μm glass seeding particles.

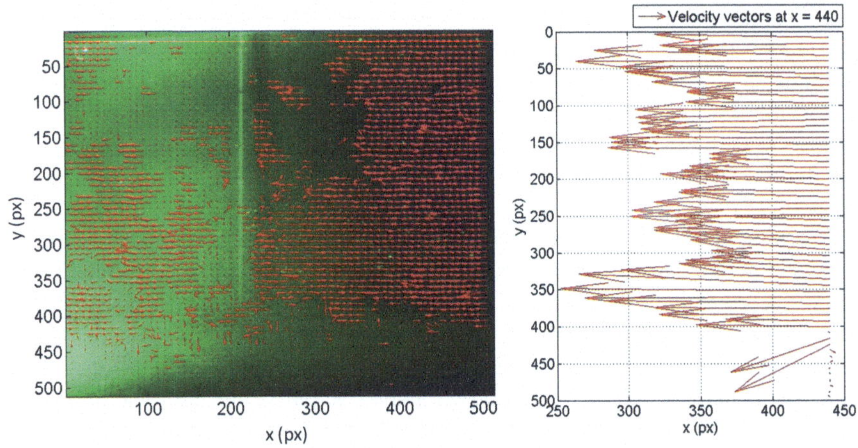

Fig. 3.18 The velocity vectors of the water with glass particle suspension (180–250 μm, 35.5$\%_{wt}$), along the axial cross-section of the small-scale rotational flow cell, determined by the particle image velocimetry method.

the upper part of the cell (1–7 cm of the flow channel's height), whereas the velocity is 0 m/s for the particles lying at the cell's bottom (0–1 cm of the flow channel's height). These values correspond to some degree to the velocity obtained using the ultrasonic TTD method for the same particle suspension, Fig. 3.12a. The results incline toward a claim that the theoretical values in Fig. 3.12c are invalid.

There is a need to obtain the average velocity profile, based on many images and applying advanced image averaging techniques, to represent the velocity with the highest accuracy. Also, to compare the results with the reference result for water with seeding particles, and to describe complex flow patterns of this dynamic suspension. Such studies were not performed in this project, however, they are recommended to be carried in future research.

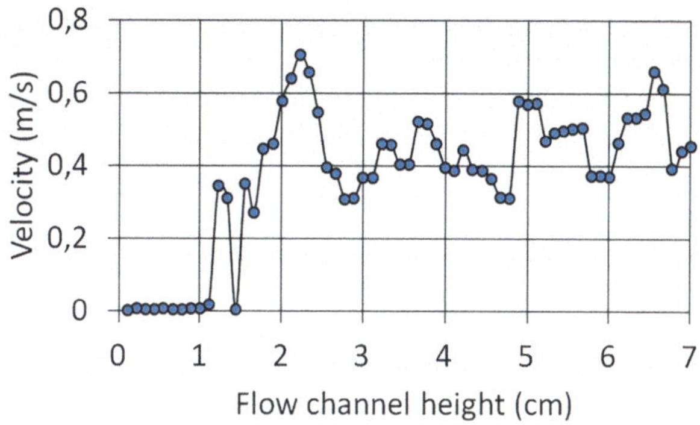

Fig. 3.19 The instantaneous, local (for x = 440 px) velocity profile of the particle suspension in Fig. 3.18, for the rotor speed of 250 RPM (3.14 m/s), determined by the particle image velocimetry method.

3.3.7 *Conclusions*

- The mean axial flow velocity distributions for water and water-particle flows were measured in the small-scale rotational test cell, for the rotor speed of 250 RPM (3.14 m/s), by the ultrasonic transit-time difference (TTD) method. Additionally, high-speed camera images were recorded for the flow velocity determination using the particle image velocimetry (PIV) method, in order to verify the ultrasonic results.
- The theoretical flow speed values (the simple Couette flow) for pure water, range from 3.14 m/s (the cell's top) to 0 m/s (the cell's bottom), Fig. 3.7d. They are much higher than the flow speed values measured by the TTD and PIV methods.
- The flow speed values of the water-glass particle suspension (180–250 μm, $35.5\%_{wt}$) measured by the ultrasonic TTD method, vary between 0.32 m/s (at the cell's top) and 0.1 m/s (at the cell's bottom), Fig. 3.12a. Similar values were obtained for pure water, however, they do not clearly decrease toward the cell's bottom. The measured flow speed values are lower than the theoretical ones, due to the secondary flow slowing down the axial flow, and a developed air ring under the rotor, reducing the transferred drag force from the rotor into the fluid.
- The instantaneous, local axial velocity profile for the water-glass particle suspension (180–250 μm, $35.5\%_{wt}$) was obtained using the laser-based PIV method, Fig. 3.19. It confirms at least to some degree, the flow velocity values determined by the ultrasonic TTD method, Fig. 3.12a. The estimated particle velocities oscillate between 0.3 and 0.7 m/s, within the cell's height range of 1–7 cm, Fig. 3.19. More advanced case studies are necessary to describe in greater detail this dynamic, heterogeneous particle suspension, using the PIV method.
- The ultrasonic testing resulted in the flow velocity values linearly decreasing toward the cell's bottom, as it was theoretically predicted, for the case of water-particle flow. However, no clear tendency was seen for the case of water flow. This is in contrary to what was expected, since the presence of the particles

should encourage the scattering of data points, by the flow fluctuations induced by the particle inertia force. This can be explained by turbulent flow structures created in by the secondary flow. A hypothesis is that the secondary flow is more pronounced for the case of pure water flow, than for the case of water-particle flow. The particle bed on the cell's bottom damped and shifted upward the kinetic energy.

- The ultrasound sensitivity measurement was performed using the ultrasound transit-time difference in the small-scale rotational test cell. The minimum detected flow speed change was 5 RPM (0.06 m/s). A preferable position for testing the sensitivity is in the upper part of the cell, due to the suspected presence of the secondary flow structures in the lower cell's part.
- The ultrasound sensitivity to water flow fluctuations was performed using the transit-time difference method in the small-scale rotational test cell. The measured average fluctuations are of the size of 2×10^{-8} s (0.054 m/s). The scattering of the data (from 3×10^{-8} to 8×10^{-8} s, from 0.1 to 0.25 m/s) indicates that the secondary flow structures were captured by the ultrasonic measurement system, which resulted in the dispersion of the average flow velocity data points.
- In the temperature test, the risk of an error introduction into the sound speed measurement by a change of the system's temperature was minimized, by the acquisition of the signals in the "with the flow" and "against the flow" directions, within the shortest time interval. However, to avoid completely the temperature influence on the sound speed measurement, the simultaneous measurements in the "with the flow" and "against the flow" directions should be performed.

3.4 The Cross-Correlation for Flow Velocity Measurement, with Data Science

3.4.1 Subproject Objectives and Challenges

3.4.1.1 The Subproject Objectives

The objective is to perform the feasibility study of the cross-correlation of ultrasound time signals, to determine the average velocity of liquid-particle flow, from backscattered echoes generated by flowing particles. The elapsed time between reflexes from the particles is used for the flow velocity measurement.

An ultrasonic Doppler flowmeter is based on a similar principle, with the transducer's face usually placed perpendicular to the flow direction. The flowmeter utilizes the frequency shift of an ultrasonic signal (the Doppler effect), when the signal is reflected by suspended particles or gas bubbles (discontinuities) in motion (Endress+Hauser 2006). This effect is utilized for the flow velocity measurement.

3.4.1.2 The Subproject Challenges

- The acquisition of a readable ultrasonic signal for the cross-correlation processing. The signal needs to contain characteristic echoes (backscattered reflexes) generated by flowing particles.
- Writing a programming code for the cross-correlation of ultrasonic signals, which assumes the flexibility in terms of defining correlation intervals and a possibility for their automatic selection.
- Creating an algorithm for the quality improvement of noisy ultrasonic signals, to obtain more accurate results.

3.4.2 An Experimental Setup

The experimental setup used for flow velocity measurement by the cross-correlation of ultrasonic signals is presented in Fig. 3.20. In the performed tests, an ultrasonic transducer was operating in the pulse-echo mode (transmitting and receiving ultrasound waves), driven by the incident spike pulse of 5 kHz PRF and 100 μJ energy. The transducer was initially placed on the external wall of the square-shaped cell, then it was immersed in fluid and placed close to the wall, with the transducer's face directed toward the flow direction. In the second configuration, a non-immersible transducer was converted into an immersible type, by putting a home-made rubber sleeve tightly on the transducer's body and leaving only the face to be in contact with fluid. After the conversion, the transducer was waterproof.

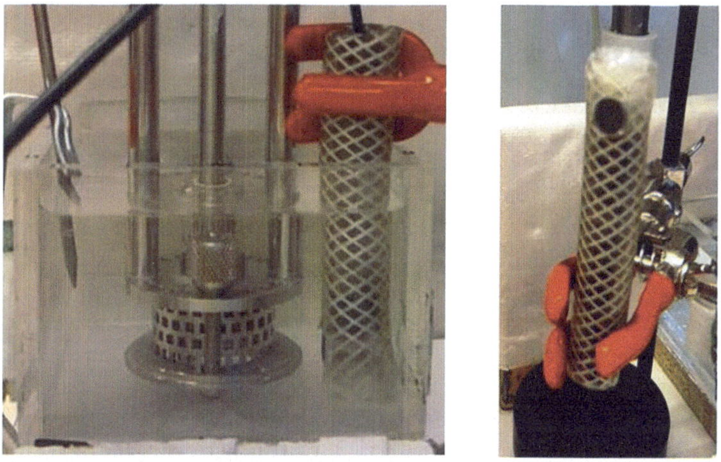

Fig. 3.20 The experimental setup used for the cross-correlation test and the ultrasonic transducer converted into an immersible type.

Tests were performed for water flow with glass particles, and the mixer velocities of 1.56–1.98 m/s were generated by the high-shear mixer. The velocities of fluid were expected to be lower than of the mixer.

3.4.3 Experimental Methods

Due to the occurrence of challenging signals for the cross-correlation, which means the signals with much noise and without any clear, characteristic backscattered echoes, the following methods for the flow velocity measurement were applied:

- Studying the water-particle flow, for the high-shear mixer velocities of 1.56–1.98 m/s, for two particle concentration groups: larger than $10\%_{wt}$ and lower than $1\%_{wt}$, using the 180–250 μm glass particles. The flow velocities created air bubbles.
- Performing the flow tests with the transducer placed outside and inside the cell.
- Cross-correlating the acquired signals in the developed MATLAB programs. Two cross-correlation functions were used: one is based on the multiplication of signals (the default method), and the other is based on the subtraction of signals (the supportive method).
- Cross-correlating the signals by picking a single interval manually, as well as by the simultaneous cross-correlation of several intervals automatically, selected by the program.
- Applying the algorithm for stretching of the signal amplitude (Sect. 3.4.5), to emphasize the most characteristic shapes in the signal, for reaching possibly the highest measurement accuracy.

3.4.4 The Cross-Correlation Principles

The principles for the cross-correlation of ultrasound time signals are the following:

- Relating the cross-correlation to taking pictures one after another, where the first picture corresponds to the first wave cycle, and the second picture corresponds to the second wave cycle.
- A representative signal for the cross-correlation consists of the wave cycles that captured echoes from flowing particles, which are "time shifted" in the next cycle.
- Using possibly flat parts of the signal.
- Extracting short time intervals from the signal for their cross-correlation.
- "Resetting" each of the intervals to the time $t = 0$, to determine the time difference.
- Using the data point numbers (the indexes) j, instead of the times t.
- Setting the maximum time shift, τ_{max}, based on the expected maximum velocity value. For example, for the maximum velocity of 2 m/s, the maximum time shift is 5.4×10^{-7} s for the pulse-echo mode.

The flow speed, U (m/s), is determined by the following equation:

$$U = \frac{\Delta L}{(1/\mathrm{PRF})} = \Delta L \cdot \mathrm{PRF} \tag{3.5}$$

where:

ΔL—the distance (m), which a particle has traveled in the time ΔT
$\Delta L = c \cdot (\Delta T/2)$ for the pulse-echo mode (m)
$\Delta L = c \cdot \Delta T$ for the through-transmission mode (m)
c—the speed of sound (m/s)
ΔT—the time difference determined by the cross-correlation method (s)
PRF—the ultrasonic pulse repetition frequency (Hz or s^{-1})

The cross-correlation functions (CCFs) are based on the signal multiplication, Eq. (3.6), and the signal subtraction methods, Eq. (3.7), for $\tau = 1, \ldots, N$.

$$\mathrm{CCF}_1(\tau) = \sum_{j=1}^{N} \left[v_1(t) \cdot v_2(t+\tau) \right] \tag{3.6}$$

$$\mathrm{CCF}_2(\tau) = \sum_{j=1}^{N} \left[\mathrm{abs}\left(v_1(t) - v_2(t+\tau) \right) \right] \tag{3.7}$$

Theoretically, the advantages of using the signal subtraction CCF, instead of the signal multiplication CCF, are the following:

- The potential of providing better resolution in the case of a very noisy signal, where its backscattered pattern is buried in the noise or has the amplitude within the noise level.
- When the amplitude of the backscattered pattern is not preserved, however, the pattern's shape is preserved.

The target of the signal multiplication CCF (the default method) is to find the first maximum, Fig. 3.21. However, for the signal with significant noise, the noisy signal's parts and not actually the backscattered parts may give the maximum CCF value. This effect could be avoided by using the signal subtraction CCF (the supportive method), where the target is to look for the first minimum, Fig. 3.22.

3.4.4.1 The Procedure for the Simultaneous Cross-Correlation of Intervals

The simultaneous cross-correlation of several intervals (the programming code is in Sect. 5.1.6) is performed according to the following procedure:

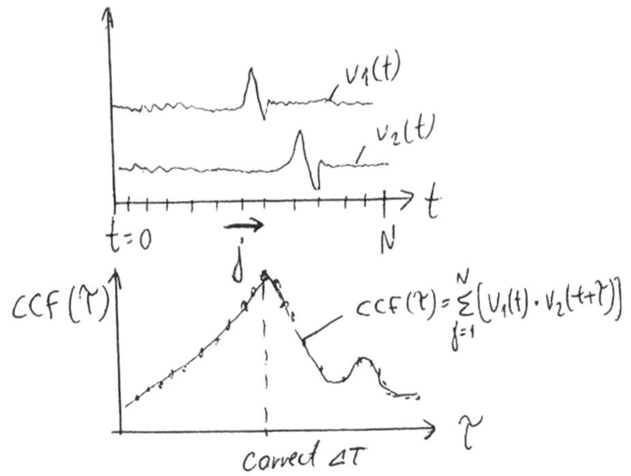

Fig. 3.21 An illustration presenting the signal multiplication cross-correlation (the default method).

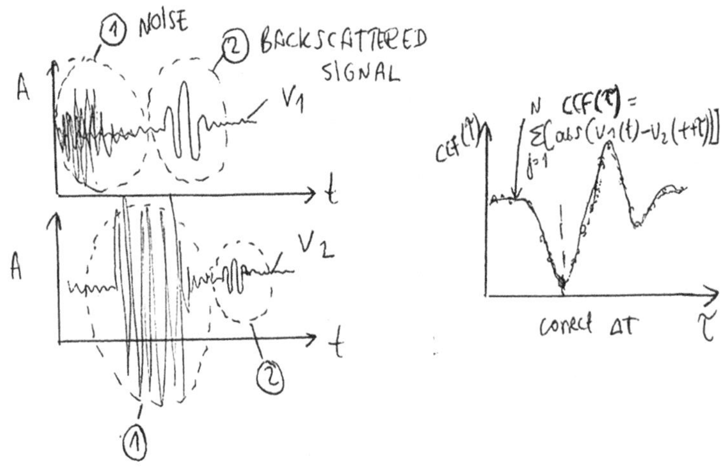

Fig. 3.22 An illustration presenting the signal subtraction cross-correlation (the supportive method).

1. Analyze only useful and possibly flat parts of a time signal. For example, the first 150 points out of the total 40,000-point interval are useful, and the rest of the signal is not useful, due to noise or spikes.
2. Apply an algorithm for stretching (magnifying) signal amplitude on the selected parts of the time signal, for example, on short 600-point intervals. Such a method will expose the characteristic shapes. The algorithm is described in Sect. 3.4.5 and the programming code is in Sect. 5.1.7.

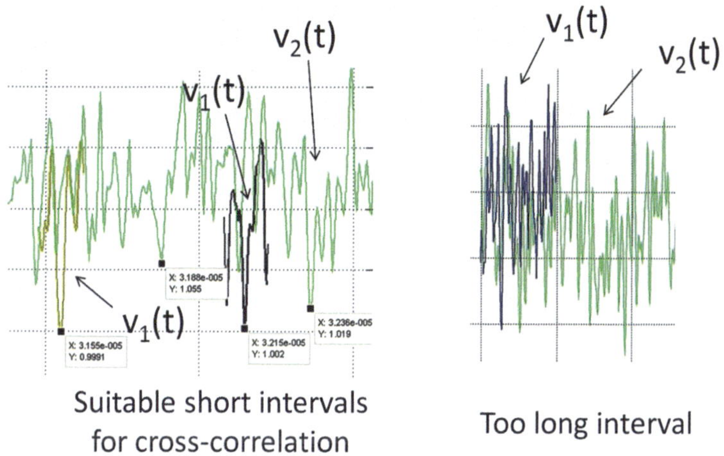

Fig. 3.23 Examples of the selected signal intervals for the cross-correlation processing, with the amplitude buried in noise.

3. Divide each selected part of the signal into short intervals, for example, 200-point intervals.
4. Apply the following algorithm:
   Perform the following “for” loop for each short 200-point interval:

   - Firstly, divide the signal into several 30-point intervals (the length and number of the intervals are flexible) and find the local minimum (or the local maximum).
   - Secondly, create new intervals around the local minimum (or the local maximum) consisting of 30 points (the lengths of the intervals are flexible). These intervals will be used for the cross-correlation, because they contain more characteristic features than the 30-point intervals, created by the signal division. It was observed that for the intervals longer than 30 points, the backscattered shapes were lost in the signal noise, Fig. 3.23.

   Perform the “for” loop for each 30-point interval:

   - Cross-correlate the intervals and plot the results.

5. Use two cross-correlation functions for the signal processing, based on the signal subtraction and the signal multiplication, Eqs. (3.6) and (3.7).

Some of the intervals that are selected automatically by the program are visually more “representative” (the first and third intervals), than the other (the second and fourth intervals), Fig. 3.24.

The cross-correlation programming code—version I (Sect. 5.1.5) was designed for the manual selection of an interval for the cross-correlation. Version II (Sect. 5.1.6) was designed for the automatic selection of multiple intervals.

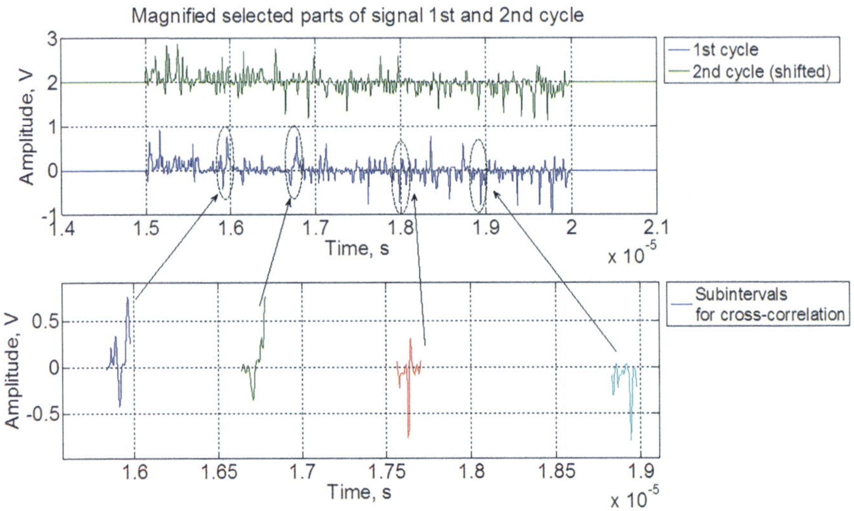

Fig. 3.24 The backscattered signal for the water-glass particle flowing suspension of the particle mass fraction lower than $1\%_{wt}$, for the mixer speed of 1.56 m/s. The 30-point intervals were automatically selected by the cross-correlation program.

3.4.5 *The Signal Amplitude Stretching Algorithm*

For the determination of flow velocity using the cross-correlation of signals, the quality of a noisy signal can be significantly improved by applying an algorithm for stretching of the amplitude. The algorithm for amplitude stretching (magnifying), suitable for time series can be found in Sect. 5.1.7. The procedure is the following:

(a) Two signal cycles are acquired, Fig. 3.25.
(b) The processing is performed by splitting the first signal cycle into short intervals, and then "zeroing out" (resetting) the amplitude, by shifting the signal by a constant for further processing, Fig. 3.26.
(c) Three amplitude segments are defined, Fig. 3.27. This needs to be done separately for the top ($A \geq 0$) and the bottom parts of the signal ($A < 0$). Then, the segments 1–3 are multiplied by the individual constants: C_1, $C_2 = 2C_1$, and $C_3 = 3C_1$, to stretch (magnify) the amplitude, with the highest impact on the most top and bottom segments.
(d) The magnified amplitude segments are then combined into the final, magnified signal, which has more characteristic features for the cross-correlation processing than the original signal, Fig. 3.28.

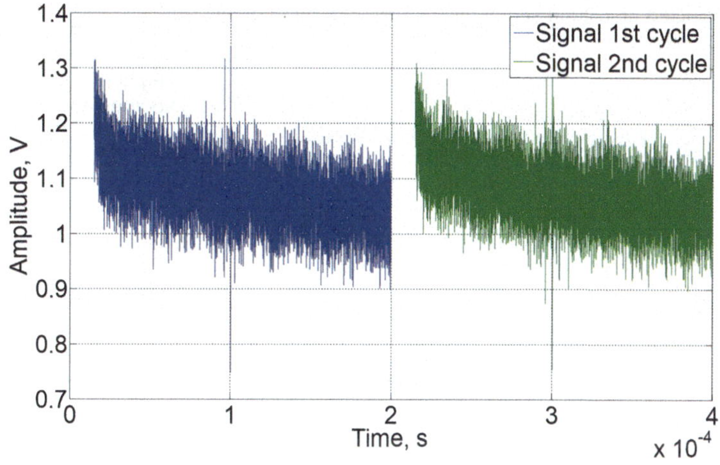

Fig. 3.25 The acquired two signal cycles for the cross-correlation process.

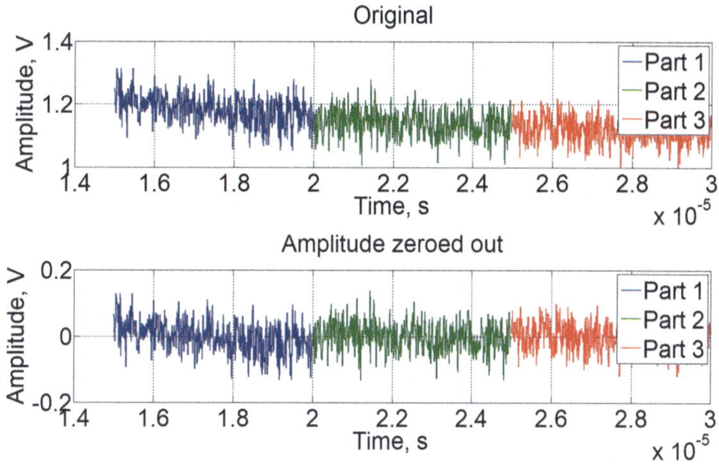

Fig. 3.26 Splitting the signal into three parts (the top figure) and "zeroing out" the amplitude (the bottom figure).

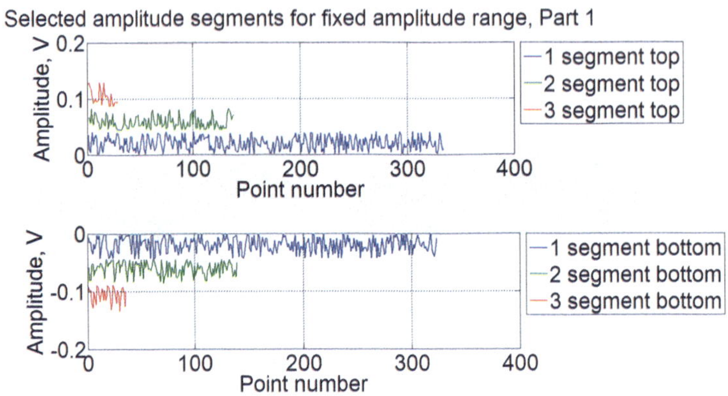

Fig. 3.27 The signal amplitude segments with positive (the top figure) and negative values (the bottom figure).

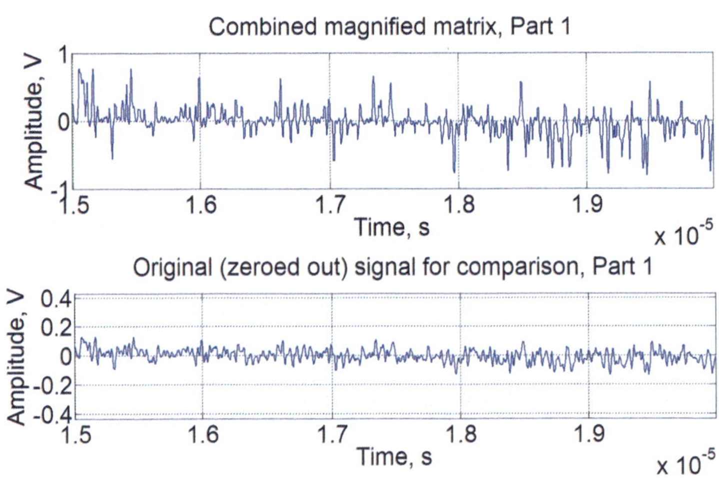

Fig. 3.28 The improved quality stretched signal for the cross-correlation processing (the top figure) and the original raw signal (the bottom figure).

3.4.6 Results

Flow velocity was determined from cross-correlation tests. The flow velocity value, determined from the original backscattered signal is 0.21 m/s, while it is 0.78 m/s from the processed signal of the improved quality. The latter signal was processed using the signal amplitude stretching algorithm (Sects. 3.4.5 and 5.1.7).

These values are compared with the reference mixer velocity of 1.56 m/s, Table 3.7. The calculated velocities are significantly lower than the reference, indicating on some inconsistency. This may be explained by the flow turbulence. The stretching algorithm applied on the original signal helped to enhance its quality for cross-correlation, Fig. 3.29. In the case of several intervals selected automatically by the program, the calculated flow velocities are from 1.05 to 3.11 m/s, compared to the reference mixer velocity of 1.56 m/s, Table 3.8 and Fig. 3.30.

Table 3.7 The fluid velocity determined by the cross-correlation method. The selected intervals for the cross-correlation were chosen manually.

Signal type	Reference mixer velocity U_{ref} (m/s)	Velocity (visually) U_{vis} (m/s)	CCF velocity U_{CCF} (m/s)
Originally acquired	1.56	0.19	0.21[a], N/A[b]
Processed using the stretching function	1.56	0.78	0.78[a], 0.78[b], 1.05[b]

[a] The signal multiplication CCF
[b] The signal subtraction CCF

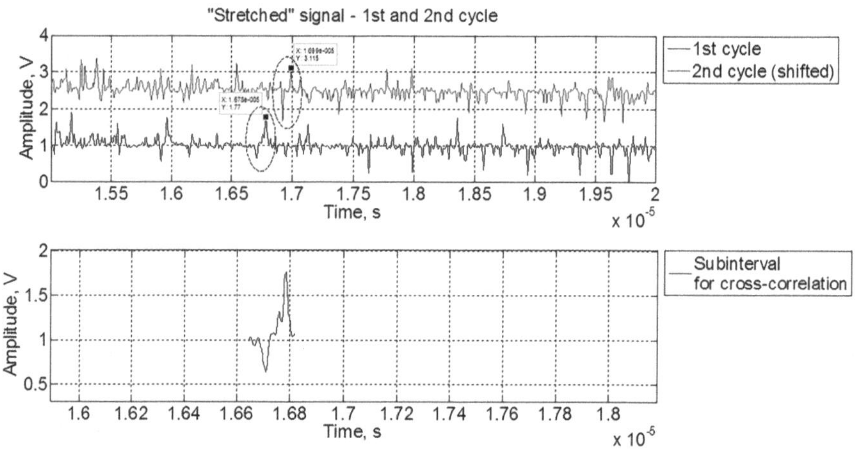

Fig. 3.29 The backscattered "processed" signals, using the signal amplitude stretching algorithm, acquired for the water-glass particle flowing suspension of the particle mass fraction lower than $1\%_{wt}$, for the mixer speed of 1.56 m/s. The selected interval (the bottom figure) is suitable for the cross-correlation processing.

Table 3.8 The fluid velocity determined by the cross-correlation of several intervals simultaneously. The intervals were selected automatically.

Signal type	Reference mixer velocity U_{ref} (m/s)	CCF velocity U_{CCF} (m/s)
Processed using the stretching function	1.56	Interval 1: 2.19[a], 2.19[b] Interval 2: 0.81[a], 1.65[b] Interval 3: 3.11[a], 3.11[b] Interval 4: 2.38[a], 1.05[b]

[a] The signal multiplication CCF
[b] The signal subtraction CCF

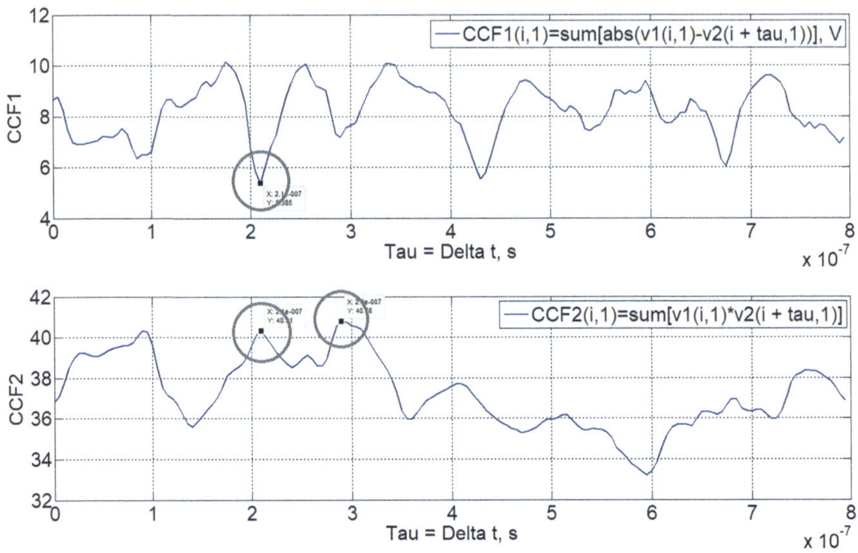

Fig. 3.30 The calculated time shift of the "processed" signal from Fig. 3.29. The CCFs are based on the subtraction (the top figure) and the multiplication (the bottom figure). The result from the signal subtraction CCF (the minimum) roughly corresponds to the results from the signal multiplication CCF (the second and third maxima).

3.4.7 Conclusions

- The cross-correlation measurement of the average velocity of the liquid-particle flow is quite challenging, however, feasible for the setup used in the test.
- The time shifts of the signals for the lower flow speeds of 1.56–1.61 m/s are possible to determine by a visual analysis, however, not possible for the higher flow speeds of 1.71–1.98 m/s, due to the turbulence causing the dampened reflected echoes.
- A standard technique for the cross-correlation of signals is based on the signal multiplication CCF. An additional, supportive technique is based on the signal subtraction CCF and is useful for noisy signals.
- The short intervals of around 30 points are optimal for the cross-correlation processing, due to usually containing a single, desirable characteristic feature. Longer intervals tend to be noisy and challenging for the processing.
- The time shift of the signal based on the visual analysis was roughly confirmed by the average time shift determined by the program. The automated cross-correlation method needs high measurement statistics at this stage. The cross-correlation methodology for liquid-particle flow is working, however, it needs quality improvements.
- The signal amplitude stretching algorithm helps to identify the time shifts of the measured signals more easily, due to exposing their backscattered reflexes. The visual analysis of the original signals is possible, however, it is quite difficult

due to the backscattered reflexes buried in the noise level. Therefore, the algorithm significantly improved the measurement accuracy.
- The simultaneous cross-correlation of several intervals selected automatically by the program, based on finding the local minimum of the signal and building small intervals around it, brought only partly satisfying results, as some of the intervals were not representative for the cross-correlation. This issue could be further improved. However, some of the selected intervals were well-selected for this experimental-digital measurement. Increasing the number of intervals would result in better statistics and therefore more accurate results.
- The test samples of the particle mass fractions lower than $1\%_{wt}$ produced the best signal quality for the cross-correlation. The test samples containing air bubbles, and the particle mass fractions higher than $10\%_{wt}$ resulted in noisy signals. They did not contain useful characteristic echoes due to too much sound scattering.
- Better signal quality was obtained using the ultrasonic transducer immersed in water, than using the transducer externally mounted on the cell's wall, due to the damped signal energy in the latter case.
- Better results are expected to be obtained in an U-shaped flow system with long and straight sections, rather than using the current test setup, which generated turbulent flow structures (vortexes) for the mixer speeds of 1.56–1.98 m/s. However, the developed methodology and programming codes have already generated the value at this stage of the research.

3.5 Acoustic Streaming with the Particle Image Velocimetry

3.5.1 Subproject Objectives and Challenges

3.5.1.1 The Subproject Objectives

- To visualize the acoustic streaming in Newtonian and non-Newtonian liquids, complemented by the particle image velocimetry (PIV) technique.
- To characterize the interactions of ultrasounds with liquid in a closed system.
- To determine velocity profiles using the PIV method.

3.5.1.2 The Subproject Challenges

- Choosing an ultrasonic transducer and the proper parameters of the incident sound wave.
- Setting up laser light illuminating liquid flow in a test cell.
- Recording images during the acoustic streaming, using a high-speed camera.
- The PIV measurement of the flow velocity profile of Newtonian fluid.

3.5.2 *The Acoustic Streaming Theory*

3.5.2.1 The Acoustic Streaming Definition

- Acoustic streaming is related to steady movement of fluid. The fluid movement is driven by the absorption of high amplitude acoustic waves by the fluid. The ultrasonic transducer transfers the momentum into the medium; if the medium is fluid and therefore free to move, it will do so, due to the ultrasound waves propagating through it.
- The movement is along the beam's direction.
- It moves the mass and creates the heat transfer in situ.

3.5.2.2 The Acoustic Streaming Patterns

The study of the streaming patterns in the glycerol and CO_2 system was performed by Fogler (1970). The fluid movement due to the acoustic streaming was possible to capture, by using tracer particles. The author observed the repeating circular patterns of the acoustic streaming, symmetrically located on the sides of the transducer beam's centerline. The streaming was the most intense along the center of the beam, and the speed was about 10 cm/s.

3.5.2.3 The Acoustic Streaming Equation

The streaming velocity, U_0 (m/s), in Newtonian fluids is described by Eckart's equation (1948):

$$U_0 = \frac{\alpha W \phi}{\eta_s c} \tag{3.8}$$

where α is the sound attenuation coefficient (Np/m), W is the total power of the beam (W), ϕ is the geometric factor related to the transducer's diameter, beam's geometry and boundaries of fluid (dimensionless), η_s is the shear viscosity (Pa·s), and c is the sound speed in fluid (m/s) (Cheeke 2002; Duck et al. 1998; Wu and Nyborg 2006).

The streaming velocity depends on the balance between the viscous forces retarding the flow, η_s, and the radiation pressure driving the stream, $\alpha W/c$, in other words the momentum in liquid generated by the transducer pressure. Any absorption mechanisms can contribute to the sound attenuation, including the shear viscosity, the bulk viscosity, and the relaxation. The attenuation due to the viscosity is described by Stokes' equation, Eq. (2.16), Sect. 2.5.4. There are also other factors that influence the sound attenuation. These are the sound loss mechanisms, according to the superposition theory, described in Sect. 2.5.3. The streaming velocity is dependent on the sound frequency, due to the sound attenuation dependence on the frequency (Figs. 2.22 and 2.23, Sect. 2.3.2).

It has been a challenge to find an equation for the acoustic streaming in non-Newtonian fluid, and a model for the non-Newtonian fluid viscosity dependent on the shear rate.

3.5.3 Example Applications

Example applications of acoustic streaming are the following:

- Non-invasive intervention purposes in the engineering and medical industries, for example, mixing of chemical solutions, ultrasonic cleaning, breaking kidney stones.
- Inducing fluid flow without applying internal mechanical force.

3.5.4 Experimental Design

Experiments were designed embracing the following elements:

- Studying the streaming effect in water, glycerine and the non-Newtonian solution of water with PAC (Polyanionic Cellulose) particles.
- Using the PIV method to visualize the induced flow.
- The streaming velocity measurement in the near field of an ultrasonic transducer for different applied voltage, using the PIV method.
- Using the tracking (seeding) particles (63–80 μm, 1050 kg/m^3) from Swiss Grilltex EMS Chemie and the color indicator Lissamine red 6B from Chroma-Gesellschaft GmbH. The seeding particles are neutrally buoyant in water.

3.5.5 An Experimental Setup

The experimental setup for the acoustic streaming with the particle image velocimetry is presented in Fig. 3.31. The setup's elements are the following:

- The Agilent waveform generator outputs the 4 MHz PRF continuous sinusoidal waves. The frequency is the same as of the transducer's natural frequency for achieving the maximum effect.
- The immersible, unfocused ultrasonic transducer from Met-Flow (the frequency of 4 MHz, the diameter of 5 mm) is immersed in the square container (the size of 20 cm), filled with fluid.
- The glass plate is used to disperse the focused laser beam, which illuminates the measurement section with uniformly distributed light.
- The laser system consists of a diode, an incorporated power supply and a controller. A collimator is placed between the laser and glass plate to narrow the beam.

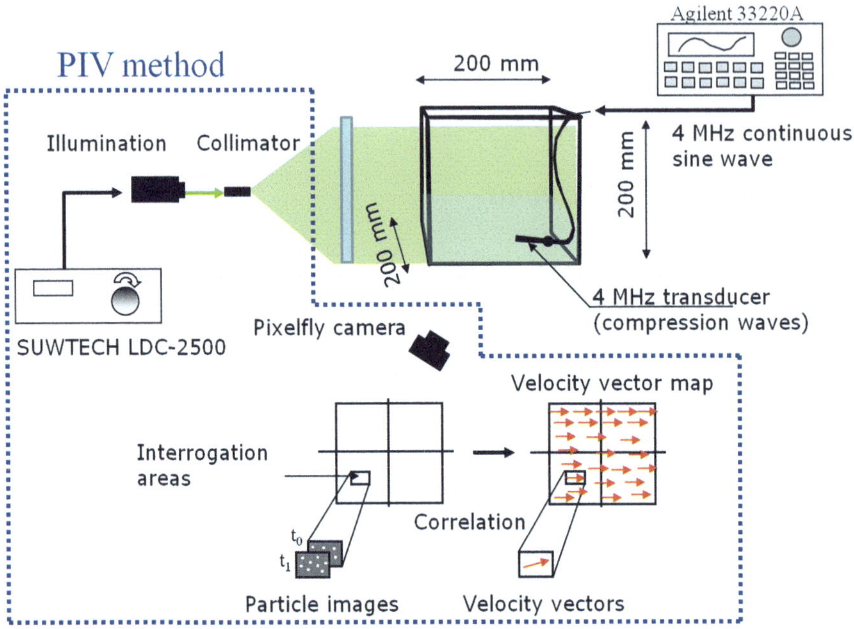

Fig. 3.31 The experimental setup used for the acoustic streaming with the particle image velocimetry. (Upgraded, originally from H.A. Rabenjafimanantsoa, B.M. Wrobel, R.W. Time, *Annual Transactions of the Nordic Rheology Society* 17 (2009)).

- The Cooke Corporation high-speed camera with the 1392 by 1024 px resolution, the frame rate of 2000 frames/s and the exposure time from 1/100,000 to 10 s, was used for the acquisition of videos and images.
- The acquired images were post-processed in DynamicStudio (from Dantec Dynamics A/S).

The particle image velocimetry method is based on the cross-correlation of two recorded video images, captured using the high-speed camera. These two images are divided into small interrogation areas (subwindows), for example, of the size of 32 by 32 px. During the time interval Δt, which is the time between two consecutive images, a single particle has moved by the displacement Δs. The absolute velocity is computed as $\Delta s/\Delta t$ and the velocity vector is along Δs. Two images are cross-correlated (in other words superposed), and the velocity vectors are calculated. For example, the time between two images taken for the water test was 65.5 ms.

3.5.6 *Results*

Results from the acoustic streaming tests are presented in Figs. 3.32, 3.33, 3.34, 3.35, and 3.36. The acoustic streaming in glycerine is shown in Fig. 3.32. The direction of the ultrasonic beam is shown by the dashed line. The color indicator was

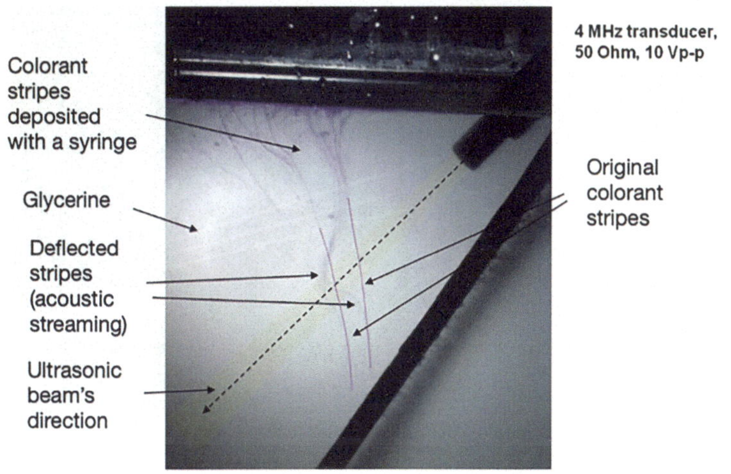

Fig. 3.32 The visualization of the acoustic streaming in glycerine. (Upgraded, originally from H.A. Rabenjafimanantsoa, B.M. Wrobel, R.W. Time, *Annual Transactions of the Nordic Rheology Society* 17 (2009)).

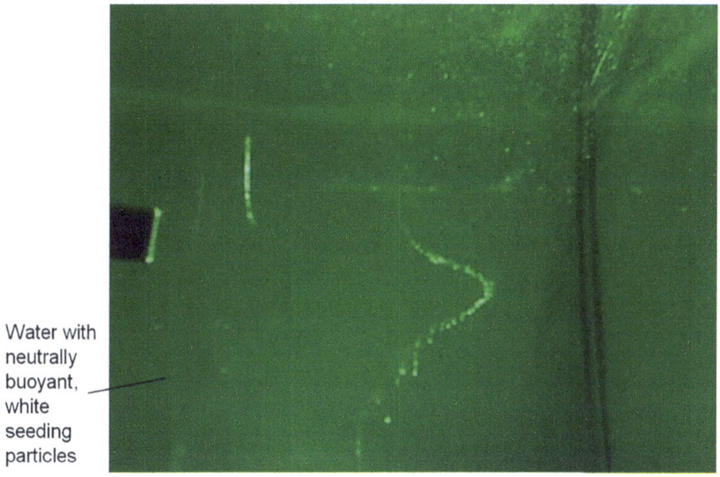

Fig. 3.33 The visualization of the acoustic streaming in water.

injected using a syringe, just below the glycerine surface. The stripes were deformed by the acoustic pressure. The original stripes are marked by the red dotted lines. The charging voltage was 10 Vpp (peak-to-peak). Two color indicator streaming traces are present; however, the streaming is more visible for one on the right-hand side. The acoustic streaming in water with naturally buoyant white seeding particles is presented in Fig. 3.33. The particles moved due to the acoustic pressure. The maximum streaming velocity is the highest in the place, where the beam intensity is the

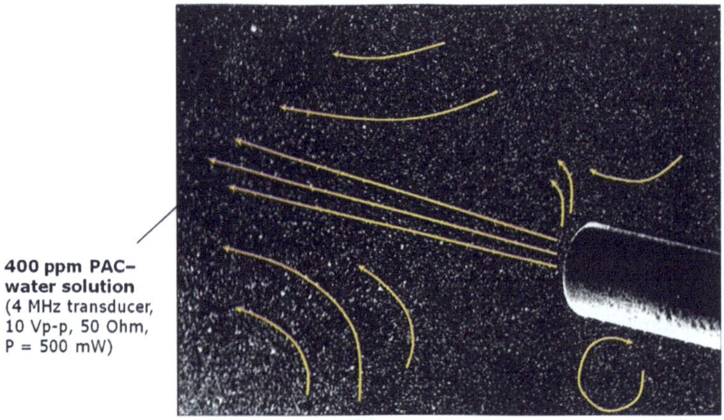

Fig. 3.34 The streaming pattern for the 400 ppm PAC (Polyanionic Cellulose) polymer particles dissolved in water.

highest. An image from the video recorded for the 400 ppm PAC Polyanionic Cellulose particles dissolved in water, is presented in Fig. 3.34. The applied voltage was 10 Vpp. No seeding particles were necessary in this case, because the polymer particles functioned as the tracer particles. The particles moved mainly in the direction along the beam, and eddies were also created in the region close to the transducer. It was a challenge to generate the acoustic streaming effect in highly viscous non-Newtonian fluid, due to its great sound attenuation properties. The arrows show the streaming pattern obtained from the recording.

Figure 3.35 shows the velocity vector map with the velocity vectors represented by the arrows, resulting from the PIV cross-correlation of the recorded images. The velocity values along the line perpendicular to the beam's centerline were extracted. The final absolute velocity equals to the length of the velocity vector and is calculated using Pythagoras' equation. The velocity vector consists of the U and V components. The vector component U is in the horizontal direction and V is in the vertical direction. The length of the velocity vector is calculated from the equation: $L = (U^2 + V^2)^{1/2}$.

The acoustic streaming velocity profile for water is presented in Fig. 3.36. The x-axis corresponds to the distance along the representative line. The y-axis corresponds to the velocity vector length. The streaming velocity increases with the increasing voltage. The maximum absolute velocity is 1.4 mm/s at the distance of 9.8 mm, obtained for the voltage of 7.5 V. This is the position of the transducer's beam centerline, where the streaming intensity is the highest. The velocity profiles are relatively symmetrical.

More information can be found in the work by Rabenjafimanantsoa et al. (2009). The programming code for calculation of the PIV flow velocity profiles is in Sect. 5.2.1.

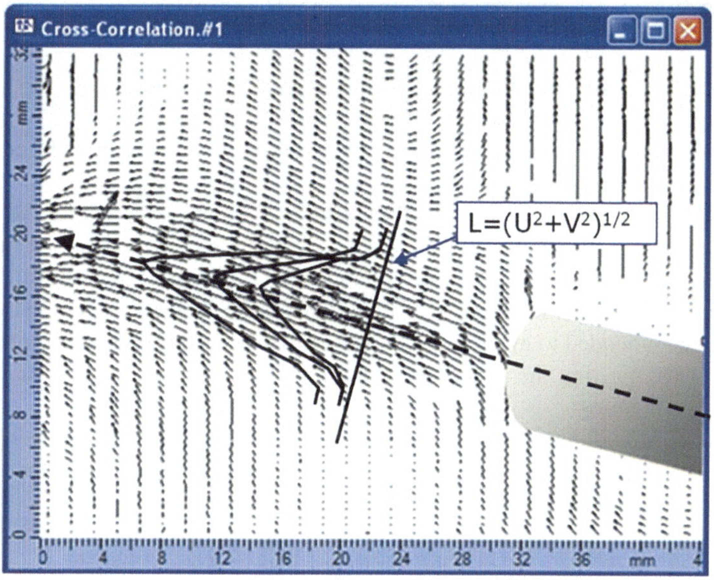

Fig. 3.35 The velocity vector map created using the particle image velocimetry (PIV) method. (Upgraded, originally from H.A. Rabenjafimanantsoa, B.M. Wrobel, R.W. Time, *Annual Transactions of the Nordic Rheology Society* 17 (2009)).

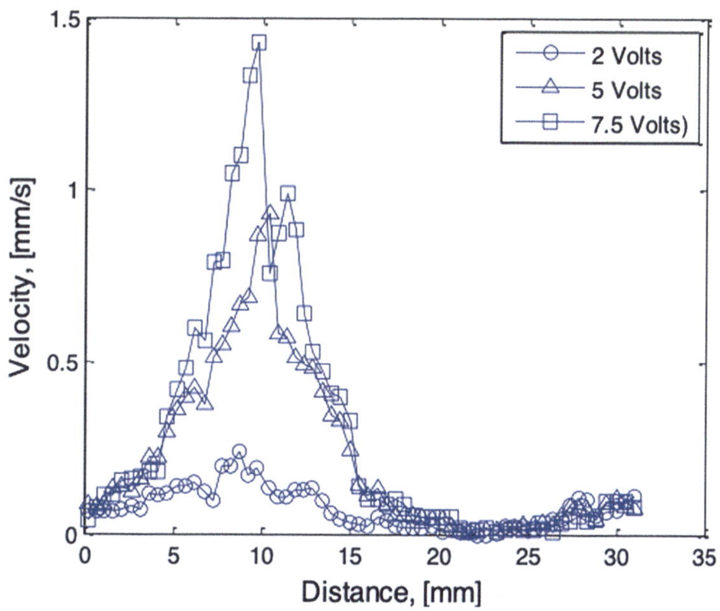

Fig. 3.36 The streaming velocity profiles for water, dependent on applied voltage. (H.A. Rabenjafimanantsoa, B.M. Wrobel, R.W. Time, *Annual Transactions of the Nordic Rheology Society* 17 (2009)).

3.5.7 Conclusions

- The acoustic streaming fluid velocities increase with the increasing applied voltage.
- The maximum peak velocity is along the axial position of the transducer, where the beam energy is the highest.
- The particle image velocimetry (PIV) technique is a valuable tool for the visualization and quantification of the acoustic streaming within the range of mm/s.
- Continuous ultrasonic waves are recommended to visualize acoustic streaming. Wave energy higher than 1 mW is preferable for the acoustic streaming in fluids.
- It is recommended to avoid standing waves, which could be created by the incident wave meeting the echoes from the opposing cell's walls, to ensure high-quality streaming.
- Acoustic streaming with the PIV technique could be an accurate tool for the measurement of fluid rheology in situ. It has been a challenge to find theories for the acoustic streaming in non-Newtonian fluid flow.

Competing Interests The author has no conflicts of interest to declare that are relevant to the content of this chapter. Part of the source material was reprinted from Annual Transactions of the Nordic Rheology Society, 17, H.A. Rabenjafimanantsoa, B.M. Wrobel, R.W. Time, "PIV visualization of acoustic streaming in non-Newtonian fluid", 2009, with permission from the Nordic Rheology Society.

References

J.D.N. Cheeke, *Fundamentals and Applications of Ultrasonic Waves* (CDC Press, Taylor & Francis Group, 2002). http://students.aiu.edu/submissions/profiles/resources/onlineBook/Z9G8f3_Fundamentals_and_Applications_of_Ultrasonic_Waves-2.pdf

F.A. Duck, A.C. Baker, H.C. Starritt, *Ultrasound in Medicine* (CRC Press, 1998), p. 2019

Endress+Hauser, *Flow Handbook. A Practical Guide: Measurement Technologies—Applications—Solutions* (Endress+Hauser Flowtec AG, Reinach, Basel, 2006)

S. Fogler, *Effect of Ultrasonic Waves on Biological Mass Transport*. Progress report. ORA Project 01319. The University of Michigan, College of Engineering, Department of Chemical and Metallurgical Engineering, 1970

B.R. Munson, D.F. Young, T.H. Okiishi, *Fundamentals of Fluid Mechanics*, 5th edn. (John Wiley & Sons, Inc., Hoboken, NJ, 2006)

H.A. Rabenjafimanantsoa, B.M. Wrobel, R.W. Time, PIV visualization of acoustic streaming in non-Newtonian fluid. Annu. Trans. Nordic Rheol. Soc. **17**, 199 (2009)

B.M. Wrobel, R.W. Time, Ultrasonic measurement and characterization of a low concentration system of solid particles in liquid, in high shear flow. Appl. Acoust. **73**(2), 117–131 (2012). https://dx.doi.org/10.1016/j.apacoust.2011.07.006

J. Wu, W. Nyborg, *Emerging Therapeutic Ultrasound* (World Scientific, Hackensack, NJ, 2006)

Chapter 4
Concluding Remarks

4.1 Main Achievements

4.1.1 The Developed Ultrasonic Methodology for Flowing Suspensions

- The selection of the hardware and software elements of the ultrasonic experimental setup for liquid-particle flow measurement.
- Tested the instrumentation in terms of the measurement sensitivity and limitations (Wrobel and Time 2011).
- Developed the principles for successful signal acquisition, comprising the concepts of an ultrasonic wave design, signal acquisition parameters, signal averaging techniques (Wrobel and Time 2011; Wrobel 2012), triggering, and wave interference (Wrobel and Time 2011).
- Studied the wave design aspects for liquid-particle measurement, resulting in the definition of the suitable incident ultrasonic pulse.
- Studied the sound resonances occurring in the measurement cell and the cell's walls (Wrobel and Time 2011).

4.1.2 The Practical Applications of the Developed Methodology

- The successful ultrasonic measurement of low concentration water-glass particle flow, by using the improved pulsed broadband ultrasonic attenuation spectroscopy (Wrobel and Time 2012).
- The successful ultrasonic measurement of water-glass particle flow in high-shear, dependent on the particle concentration and flow speed (Wrobel and Time 2012).

B. Omylska, *Ultrasonic Measurement and Characterization of Liquid-Particle Flow*, https://doi.org/10.1007/978-3-032-03957-6_4

- Developed the spectroscopy method, which is suitable for the simultaneous measurement of the average concentration and flow speed, using the same ultrasonic square pulse (Wrobel and Time 2012).
- The study of the flow turbulence intensity, initiation of particle clustering, and detection of air bubbles as the third phase in water-solid particle flow (Wrobel and Time 2012).
- Proved the potential of the spectroscopy methodology for characterizing fluids having various physical properties.
- Proved the potential of the spectroscopy methodology for measuring the particle size of flowing liquid-particle systems, however, advanced signal interpretation is needed due to extra factors, such as the interparticle distance and flow effects (Wrobel and Time 2012).
- The suitable physical, acoustic, and flow dynamics theories were used for the interpretation of the experimental results (Wrobel and Time 2012).
- The superposition theory by Dukhin and Goetz (1996, 2002, 2017) could be extended to one more element, α_{turb}, which is the sound loss due to the flow turbulence intensity, $\varepsilon(Re)$. This extra element shall be taken into consideration for a complete description of the total sound attenuation in colloids. The degrees of dominating sound loss mechanisms will be different for different applications.
- The determination of the measurement and noise ranges in the sound frequency spectra, used for the acoustic spectroscopy measurement of liquid-particle flow (Wrobel and Time 2012).
- The studies of the sound speed and sound attenuation measurement uncertainties, which resulted in the listing of different impactful factors (Wrobel and Time 2012).
- Developed several numerical simulations, which supported the interpretation of the project's experimental data.
- For better understanding of the experimental results, the visualization of the sound wave propagation and sound pressure distribution in water and water containing solid particles was performed by 4D finite element modeling in COMSOL Multiphysics.
- Determined flow velocity by the cross-correlation of ultrasonic signals backscattered from solid particles. The method proved to be feasible; however, it requires further improvements in terms of the measurement accuracy.
- Developed the signal amplitude stretching program (the algorithm), which improves the measurement accuracy of the cross-correlation method.

4.1.3 The Small-Scale Rotational Flow Cell Studies

- Designing and building the test cell with its annular flow channel, for the measurement of the velocity and concentration profiles of liquid-particle flow, using ultrasound and laser-based methods.

- Determination of the theoretical flow characteristics using the simple Couette flow theory.
- Characterization of flow regimes created in the test cell. The types of formed liquid-particle flow regimes and structures are influenced by the particle size and flow velocity. For example, a well-distributed aqueous suspension or symmetrical dunes were formed, depending on the chosen parameters.
- Performed the measurements of the flow velocity distribution for flowing liquid and a liquid-particle suspension (the mean axial flow velocity along the cell's height), using the ultrasonic transit-time difference method.
- Measured the ultrasonic measurement sensitivity to flow fluctuations.
- Measured the ultrasonic measurement sensitivity to various flow speeds.
- Determined the velocity profile of a flowing liquid-particle suspension using the particle image velocimetry method, which validated the flow velocities measured by the ultrasonic transit-time difference method.

4.1.4 The Acoustic Streaming Study

- Successfully visualized the acoustic streaming in Newtonian and non-Newtonian fluids (Rabenjafimanantsoa et al. 2009; Wrobel 2012).
- Determined the velocity profiles for water, using the particle image velocimetry method (Rabenjafimanantsoa et al. 2009).
- The acoustic streaming equation enabled describing the dependences between the streaming velocity, the rheology of fluid, and the sound attenuation.

4.1.5 Data Science

A number of programming codes developed in MATLAB, complemented digitally the experimental work.

4.1.6 Project Management

- Project management methods and tools enabled the project delivery. Agile, lean and waterfall (hybrid) management methods were utilized, as well as ongoing constructive reflection. The framework involved defining the project scope, requirements and limits, daily project planning (including a roadmap), the creation of the front-end strategy, the evaluation and selection of concepts, the development of the project tasks' activities and prioritization, the current status capture and decisions for the future, the integration of the subprojects into the complex long-term project, progress reporting, the monitoring of the project

metrics, the conceptual-experimental-modeling knowledge implementation, continuous improvement, and the project delivery.
- Importance was also placed on the utilization of scientific standards, for example, ISO standards.
- Relation management and cooperation with external institutions, writing, publishing and public speaking on the subject, as well as the experience of received mentorship and given coaching were inherent to the project.

4.2 Summary

The objective of this research project was to develop the ultrasonic methodology for measurement and characterization of flow containing solid particles in online (real-time) measurement conditions.

Ultrasonic methods offered possibilities for measurement of flow dynamics parameters. Performed studies included the measurement of particle concentration and flow speed by the same type of pulse, the ultrasound sensitivity to flow turbulence intensity, the characterization of the rheological properties of fluids, the ultrasonic detection of particle clustering, and air bubbles in liquid-solid particle flow.

The improved pulsed broadband ultrasonic attenuation spectroscopy is a useful and accurate testing tool for suspensions, such as liquid-particle flow containing small diameter particles. The method is particularly useful due to using a very short pulse producing fast measurement, which is essential for studying dynamic flow systems. Additionally, the method provides a wide range of spectral frequencies, not limited to the transducer's resonance frequency. The method can be used for the simultaneous measurement of particle concentration (by the frequency spectra) and flow speed (by the sound speed measurement using the ultrasonic transit-time difference technique). This measurement can be performed using a single square pulse. The interpretation of the results is complex due to multiple factors affecting the measurements.

A number of experimental and numerical examples, together with comprehensive explanations, show the potential for the improved pulsed broadband ultrasonic attenuation spectroscopy to be a successful methodology for online measurement of liquid-particle suspensions and composite non-homogenous materials.

The project resulted in creation of the small-scale rotational test cell with its annular flow channel. It can be used for studying various liquid-particle systems, such as suspensions and dunes, the measurements of the concentration and velocity profiles using ultrasonic and laser-based methods, supported by the digital signal and image processing. The particle image velocimetry is a valuable laser-based method utilizing the image processing. This technique enables the visualization and quantification of the acoustic streaming, for very low flow velocities in the range of mm/s. The acoustic streaming with PIV could be an accurate tool for the measurement of fluid rheology in situ.

Programming codes in MATLAB and 4D finite element modeling of the acoustic fields in liquid-particle systems using COMSOL Multiphysics, complemented digitally the experimental work.

Project management methods and tools supported successful cooperation between stakeholders, for which the author is grateful, as well as the successful delivery of the project. May the curiosity of readers be instigated by this book.

4.3 Future Research Recommendations

Future research recommendations are the following:

- Testing the pulsed broadband ultrasonic attenuation spectroscopy in large-scale applications, in terms of the simultaneous measurement of concentration and velocity of liquid-solid particle flow.
- Obtaining the average velocity profile of a particle suspension in a rotational flow cell, using the particle image velocimetry technique, based on a big number of acquired images. Applying advanced image averaging techniques for increasing the measurement accuracy. Describing complex flow patterns in this type of a flow cell.
- Studying the rheological parameters of fluids in detail by means of the acoustic streaming, possibly aided by the particle image velocimetry.
- Studying different models for determination of the density and compressibility of a suspension (Eq. 2.9 in the phenomenological model by Harker and Temple 1988, Sect. 2.5.3), to define better the sound speed in the flowing suspension and the drag coefficients for the turbulent flow. It is recommended to consider these relations together with the acoustic theories for particulates (Manninen and Taivassalo 1996, Harker and Temple 1988, Sect. 2.5.3), to embrace a bigger picture, based on the already developed, innovative research.

Competing Interests The author has no conflicts of interest to declare that are relevant to the content of this chapter. Part of the source material was reprinted from: (1) Applied Acoustics, 73 (2), B.M. Wrobel, R.W. Time, "Ultrasonic measurement and characterization of a low concentration system of solid particles in liquid, in high shear flow", pp. 117–131, February 2012, (2) Applied Acoustics, 72 (6), B.M. Wrobel, R.W. Time, "Improved pulsed broadband ultrasonic spectroscopy for analysis of liquid-particle flow", pp. 324–335, May 2011, with permission from Elsevier, (3) Annual Transactions of the Nordic Rheology Society, 17, H.A. Rabenjafimanantsoa, B.M. Wrobel, R.W. Time, "PIV visualization of acoustic streaming in non-Newtonian fluid", 2009, with permission from the Nordic Rheology Society.

References

A.S. Dukhin, P.J. Goetz, Acoustic spectroscopy for concentrated polydisperse colloids with high density contrast. Langmuir **12**, 4987–4997 (1996)

A.S. Dukhin, P.J. Goetz, *Ultrasound for Characterizing Colloids. Particle Sizing, Zeta Potential, Rheology*, 1st edn. (Elsevier, New York, 2002)

A.S. Dukhin, P.J. Goetz, *Characterization of Liquids, Dispersions, Emulsions, and Porous Materials Using Ultrasound*, 3rd edn. (Elsevier, Amsterdam, 2017)

A.H. Harker, J.A.G. Temple, Velocity and attenuation of ultrasound in suspensions of particles in fluids. J. Appl. Phys. **21**, 1576–1588 (1988)

M. Manninen, V. Taivassalo, *On the Mixture Model for Multiphase Flow* (VTT Publications, Espoo, 1996)

H. Rabenjafimanantsoa, B. Wrobel, R. Time, PIV visualization of acoustic streaming in non-Newtonian fluid. Annu. Trans. Nordic Rheol. Soc. **17**, 199 (2009)

B.M. Wrobel, *Ultrasonic Measurement and Characterization of Liquid-Particle Flow*. PhD thesis no. 171, University of Stavanger, Norway, 2012

B.M. Wrobel, R.W. Time, Improved pulsed broadband ultrasonic spectroscopy for analysis of liquid-particle flow. Appl. Acoust. **72**(6), 324–335 (2011). https://dx.doi.org/10.1016/j.apacoust.2010.11.013

B.M. Wrobel, R.W. Time, Ultrasonic measurement and characterization of a low concentration system of solid particles in liquid, in high shear flow. Appl. Acoust. **73**(2), 117–131 (2012). https://dx.doi.org/10.1016/j.apacoust.2011.07.006

Chapter 5
Programs for Time Series and Images in MATLAB

5.1 Programming Codes for Time Series

5.1.1 The Frequency Sweep Signal

```
%
clear all
buffer = [];

stepsize = 1e-8              %dt between time points;
t = 0:stepsize:5e-5;         %1 sweep cycle consists of the frequencies
                             %of 1-10 MHz (50 cycles of the 1 MHz wave
                             %are in 1 sweep, with 100 points per wave
                             %cycle, 500 cycles of the 10 MHz wave are
                             %in 1 sweep with 10 points per wave cycle);
                             %5000+1 time points are in the sweep cycle;
size_t = size(t)

df = 0;
for i = 1:46                 %n = 46 for the 1-10 MHz frequency sweep;
 df;
 fr = (1*10^6)+(df*10^6); %The 1:0.2:n MHz frequencies (1, 1.2, 1.4,
                             %…, 10 MHz);
 df_m(I,:) = df;
 fr_m(I,:) = fr;
 df = df+0.2;
end
```


B. Omylska, *Ultrasonic Measurement and Characterization of Liquid-Particle Flow*, https://doi.org/10.1007/978-3-032-03957-6_5

```
%df_m
fr_m;                       %The frequency array for the sweep;
f = fr_m

w = 2*pi()*f;               %The angular frequency array for the sweep;
size_w = size(w)
help = size_w(1,1)

%Adjusting the amplitude
A = 1
for i = 1:1:help
amplA = A*sin(w(i,1)*t);
ampl_arrayA(i,:) = amplA;
%A = A+0.2;
end

SumAmplArray = sum(ampl_arrayA(1:help,:)); %The sum of the sine
                            %functions with the corresponding
                            %frequencies;
size_ampl = size(SumAmplArray)
size_ampl_help = size_ampl(1,2);
ampl = SumAmplArray;

figure(1)
subplot(3,1,1);plot(t,ampl)
title('Linear frequency sweep')
xlabel('Time (s)')
ylabel('Amplitude')
grid on

figure(1)
N = length(ampl)
Y = fft(ampl,N);            %Converting the signal to its frequency
                            %domain, for creating the frequency
                            %spectrum;
Pyy = Y.*conj(Y)/N;         %Converting the signal to its power domain
                            %(the power spectrum consists of power
                            %values for a range of frequencies);

Numpsek = 1/stepsize;       %The sampling frequency (numbers
                            %per second);
Adr = fix(N/2)
f = Numpsek*(0:Adr)/N;
```

```
subplot(3,1,2);plot(f,Pyy(1:Adr+1)) %The calculated power for
                                    %plotting the first half+1 of
                                    %points;
Sumcount = sum(Pyy(1:Adr+1))
title('Frequency spectrum')
xlabel('Frequency (Hz)')
ylabel('Amplitude')
grid on

figure(1)
subplot(3,1,3);loglog(f,Pyy(1:Adr+1))
Sumcount = sum(Pyy(1:Adr+1))
title('Frequency spectrum, logarithmic scale')
xlabel('Frequency (Hz)')
ylabel('Amplitude')
grid on
```

5.1.2 *The Sharp Square and Sines Waves with Exponential Noise*

Designing sharp square and sine waves with exponential noise. The waves are built with the following components: the sampling frequency, the number of cycles, the number of points, the amplitude, the wave frequency (the wavelength), the wave period.

```
%
buffer = [];
A = 4;
int_freq = 20*10^6
no_cycles = 20
output_freq = int_freq/no_cycles
no_points_percycle = 200
no_points_perwaveform = no_points_percycle*no_cycles;
t_stop = 1/output_freq;
T= 1/int_freq;                   %The wave period;
stepsize = t_stop/(no_points_perwaveform-1);
t = 0:stepsize:t_stop;

dt = t(1,2)-t(1,1);
sampling_freq = 1/dt
t0 = t_stop/2
sigma = (1/int_freq)*6
```

```
help = size(t);
help2 = help(1,2);

m = 10;                    %m needs to be an even number;
t = 0;

%CHOOSE THE SQUARE WAVE OR THE SINE WAVE;

%THE SINE WAVE;
%for i = 1:help2;          %The loop is necessary due to (t-t0);
%y_sq2 = A*exp((-1)*((t-t0)/sigma)^m)*sin(2*pi()*int_freq*t);
%Y_SQ2(i,1) = y_sq2
%t = t+stepsize;
%end

%Y_SQ2;
%y_sq2 = Y_SQ2;
%size_y_sq2 = size(y_sq2)

%OR;

%THE SQUARE WAVE;
for i = 1:help2;           %The loop is necessary due to (t-t0);
y_sq2 = A*exp((-1)*((t-t0)/sigma)^m)*sign(sin(2*pi()*int_
freq*t));
Y_SQ2(i,1) = y_sq2;
t = t+stepsize;
end

Y_SQ2;
y_sq2 = Y_SQ2;
size_y_sq2 = size(y_sq2)

%THE NOISE;
t = 0:stepsize:t_stop;
noise = 2*(sin(2*pi()*int_freq*0.5*t));
%noise = 2*(sin(2*pi()*int_freq*0.5*t))+2*(sin(2*pi()
*int_freq*2*t));

figure(1)
plot(t,noise,'-*')
grid on
```

```
t = 0:stepsize:t_stop;
time = t;
ampl = y_sq2';                %The transformed amplitude matrix;

figure(2)
subplot(2,1,1);plot(time,ampl,'-*')
xlabel('Time (s)')
ylabel('Amplitude')
title([ num2str(int_freq),'Hz wave'])
grid on

N = length(ampl);
Y = fft(ampl,N);              %Converting to the frequency domain
                              %(the frequency spectrum);
Pyy = Y.*conj(Y)/N;           %Converting to the power domain
                              %(the power spectrum - power values
                              %at various frequencies);
Numpsek = 1/stepsize;         %The sampling frequency (numbers per
                              %second);
Adr = fix(N/2);
f = Numpsek*(0:Adr)/N;
Sumcount = sum(Pyy(1:Adr+1));
G_in_ampl = Pyy(1:Adr+1);     %The amplitudes in the frequency
                              %spectrum;
size_G_in_ampl = size(G_in_ampl)

figure(2)
subplot(2,1,2);loglog(f,Pyy(1:Adr+1))
title('Frequency spectrum, logarithmic scale')
xlabel('Frequency (Hz)')
ylabel('Amplitude')
grid on

%THE SQUARE WAVE WITH THE NOISE;
y_sq2_noise = y_sq2'+noise;
time = t;
ampl_noise = y_sq2_noise;

figure(3)
subplot(2,1,1);plot(time,ampl_noise,'-*')
grid on
xlabel('Time (s)')
ylabel('Amplitude')
title([ num2str(int_freq),'Hz wave with noise'])
```

```
N = length(ampl_noise);
Y = fft(ampl_noise,N);
Pyy = Y.*conj(Y)/N;
Numpsek = 1/stepsize;
Adr = fix(N/2);
f = Numpsek*(0:Adr)/N;
Sumcount = sum(Pyy(1:Adr+1));
G_out_ampl = Pyy(1:Adr+1);
size_G_out_ampl = size(G_out_ampl)

figure(3)
subplot(2,1,2);loglog(f,Pyy(1:Adr+1))
title('Frequency spectrum of medium response, logarithmic scale')
xlabel('Frequency (Hz)')
ylabel('Amplitude')
grid on

%H = abs(G_out_ampl-G_in_ampl);
H = (G_out_ampl-G_in_ampl);

figure(4)
loglog(f,H)
title('Frequency spectrum difference, logarithmic scale')
xlabel('Frequency (Hz)')
ylabel('Amplitude')
grid on
```

5.1.3 The Ensemble Average of Multiple Frequency Spectra

An ensemble average is a method that is commonly used for signal averaging. The method described here was used, for example, for the calculation of the average frequency spectrum, based on the acquired time-dependent experimental data in multiple data files.

```
%
format long
clear
buffer = [];
path_s = 'F:\...\Data_set';
```

```
path(path,path_s);
name1 ='Reference_square4400p_200MSs_10avg';
name2 ='.lvm';

freq = 200*10^6;            %The input sampling frequency;

%Processing;
n = 10;                     %Load n number of files;
for i = 1:n;
filename = [name1,name2];
A =
textread(filename,'',-1,'delimiter','\t','headerlines',0,'emptyval
ue',NaN);
time = A(:,1)*(1/freq);
ampl = A(:,2);
AMPL_MATRIX(:,i) = ampl;
end

AMPL_MATRIX;
help = size(AMPL_MATRIX);

no_points = help(1,1);

for j = 1:no_points;
av_ampl = mean (AMPL_MATRIX(j,:));
AV_AMPL_MATRIX (j,:) = av_ampl;
end

display (['Amplitude mean from',num2str(n),'acquisitions of',
num2str(no_points),'data points:']);
AV_AMPL_MATRIX;

%Plotting;
figure(1)
plot(time,AV_AMPL_MATRIX)
xlabel('Time (s)')
ylabel('Amplitude (V)')
%legend('water')
grid on

ampl = AV_AMPL_MATRIX;
size_processed = size(AV_AMPL_MATRIX)
N = length(ampl)
Y = fft(ampl,N);            %Converting to the frequency domain
```

```
                             %(the frequency spectrum);
Pyy = Y.*conj(Y)/N;          %Converting to the power domain
                             %(the power spectrum - power values
                             %at various frequencies);

figure(2)
Numpsek = freq;              %The sampling frequency (numbers per
                             %second);
Adr = fix(N/2);
f = Numpsek*(0:Adr)/N;
loglog(f,Pyy(1:Adr+1))       %The calculated power for plotting the first
                             %half+1 of points;
Sumcount = sum(Pyy(1:Adr+1));
title('Frequency spectrum')
xlabel('Frequency (Hz)')
ylabel('Amplitude')
grid on

%Saving the processed data into a file;
%time = A(:,1);
%name3 = '_PROCESSED';
%savefile = [name1,name3,name2];
%save(savefile,'time','AV_AMPL_MATRIX','-ascii');
%clear all
```

5.1.4 The Running Average of a Wave Signal

A running average is a method that is commonly used for signal averaging and smoothing. The method described here was used, for example, for the processing of the signals from the ultrasonic transducer's performance test. The used signals in the test were sine waves at various frequencies. Before averaging for a big data file, testing it with a small number of data points (setting m to a small number) is the advised and useful rule of thumb. The calculations are time consuming and take a vast amount of computer memory, thus this will ensure the quality averaging.

The program's workflow is the following:

1. The algorithm is suitable for averaging, for example, noisy sine-shaped signals with evenly distributed noise over time. Other signals than the sine signals, can also be averaged, however, the quality verification needs to be performed for checking, if the curve smoothness condition needs or does not need to be modified to a certain degree.
2. The program averages and plots the signal.

3. Variables that need to be specified are the following:
   - The file name and the path, *p*.
   - The number of analyzed data points, *m*, from the file.
   - The sampling frequency, *freq*.
   - The number of points for averaging, *N*.
   - The chosen number of points that will be used for checking the slope of a curve (*diff_loop_stop* in the code, the curve smoothing part).
   - The number of the wave cycles acquired in one acquisition window, *nc*.
   - The chosen number of extra iterations, *Const*, after the critical smoothing condition is satisfied.
   - The chosen number of other extra iterations to run the loop, *Const2*, after visually checking the output.
4. The program outputs the plots and saves the data file with the averaged amplitude and time data.

The methodology is to calculate one average point from 20 consecutive data points, *N*, and then another average point from the set of the next 20 consecutive data points, etc. All 19 points will always overlap from the previous dataset and one point will be new. Averaging at the step $n + 1$ is based on the averaged amplitude from the previous run at the step n, therefore the file size (with the average amplitude and average time) decreases, as the number of the iterations for averaging increases. The instantaneous standard deviations of the original and averaged signals are calculated from the individual sets of 20 consecutive points and are plotted in a graph as the function of time. The average standard deviations from all points are given for the quantification of both cases.

The averaging procedure is repeated for a certain number of times, until one of the following two conditions are fulfilled:

(a) The consecutive data points gradually decrease in a selected (falling) part of the curve.

This part is determined through the first maximum and minimum. The maximum is found in the small interval of 1–2 cycles from the average amplitude data. The minimum is found by locating the change of the curve slope. Afterwards, the amplitude differences between two neighboring points are calculated for checking the curve monotonicity. If some of the amplitude differences are negative (*Z1* and *Z2* consist of both 1 and 0), then running the averaging loop is continued, until all the differences are positive. If all the differences are positive (*Z1* consists only of 1 and *Z2* only of 0), then the number of extra iterations is run, defined by *Const*. Afterwards, if the smoothing condition is fulfilled, the main "while" loop is paused.

(b) If the plotted averaged curve looks smooth, the iteration process can be stopped by pressing "Ctrl+C". However, by pressing any key, 5 more iterations or the other number defined by $G = Const - Const2$ can be performed. This is repeated until a user decides to stop the averaging loop.

The program is automatically stopped, if the number of data points in the input file is lower, than the assigned number of data points for averaging ($d < 0$). The iteration loop in the program runs only for $d > 0$. In the case that *Z1* consists only of 1, and *Z2* only of 0, then *Smoothing_fac* = 1 (1 is "true" and *G* becomes 1), otherwise, *Smoothing_fac* = 0 (0 is "false" and *G* is still 0). *Const2* is input externally.

```
%
buffer = [];
clear all
format long e
p = 'F:\...';
path(path,p)

  pre_0 ='100kHz';
  % pre_0 ='200kHz';
  % pre_0 ='400kHz';
  % pre_0 ='1MHz';
  % pre_0 ='2MHz';
  % pre_0 ='3MHz';
  % pre_0 ='4MHz';
  % pre_0 ='5MHz';
  % pre_0 ='6MHz';
  % pre_0 ='7MHz';
  % pre_0 ='8MHz';
  % pre_0 ='9MHz';
  % pre_0 ='10MHz';
  % pre_0 ='15MHz';
  % pre_0 ='20MHz';

name = '_sine_10Vpp_TTwater_40cyc.lvm';
filename0 = [pre_0,name] %For example:
                          %'100kHz_sine_10Vpp_TTwater_40cyc.lvm';

A0 = textread(filename0,'',-1,'delimiter','\t','headerlines', 0,
'emptyvalue',NaN);
freq = 200*10^6;

%Choose the number of points, m. The output file size is regulated by
%setting m. From the test: m = 80 for the 100 kHz file;
%m = 80;
%m = 8000;

%OR

m0 = size(A0);            %The full file size;
m = m0(1,1);
```

```
time0 = A0(1:m,1)*(1/freq);
ampl0 = A0(1:m,2);
clear A0;

%Input the number of cycles, nc0;
nc0 = 40;                    %Input the number of cycles in the original
                             %input file (see also: Help2b);
nc = nc0-14;                 %For the 100 kHz file: 80000/26 = 3000,
                             %3000 points is 1.5 cycles; this will give
                             %a short time window to find the local
                             %maximum;

%The testing value;
%nc0 = 4;
%nc = nc0-1.4;

%The initial conditions;
ampl0_ind = ampl0;           %For example (500,1);
time0_ind = time0;           %For example (500,1);

%THE DATA FOR THE MAIN LOOP;
f0 = size(ampl0_ind);
b = f0(1,1);                 %For example, 500;

N = 20;                      %Input the number of points for averaging
d = b-N+1;                   %Iterations i until 480+1;

%The subprogram calculating the standard deviation of the
%original signal for plotting;


for i = 1:d                  %'FOR' LOOP 1;
     y0(:,i) = [ampl0_ind(i,1) ampl0_ind(i+1,1) ampl0_ind(i+2,1)
     ampl0_ind(i+3,1) ampl0_ind(i+4,1) ampl0_ind(i+5,1)
     ampl0_ind(i+6,1) ampl0_ind(i+7,1) ampl0_ind(i+8,1)
     ampl0_ind(i+9,1) ampl0_ind(i+10,1) ampl0_ind(i+11,1)
     ampl0_ind(i+12,1) ampl0_ind(i+13,1) ampl0_ind(i+14,1)
     ampl0_ind(i+15,1) ampl0_ind(i+16,1) ampl0_ind(i+17,1)
     ampl0_ind(i+18,1) ampl0_ind(i+19,1)];
     %t0 (:,1)... t0(:,481)
     t0(:,i) = [time0_ind(i,1) time0_ind(i+1,1) time0_ind(i+2,1)
```

```
    time0_ind(i+3,1) time0_ind(i+4,1) time0_ind(i+5,1)
    time0_ind(i+6,1) time0_ind(i+7,1) time0_ind(i+8,1)
    time0_ind(i+9,1) time0_ind(i+10,1) time0_ind(i+11,1)
    time0_ind(i+12,1) time0_ind(i+13,1) time0_ind(i+14,1)
    time0_ind(i+15,1) time0_ind(i+16,1) time0_ind(i+17,1)
    time0_ind(i+18,1) time0_ind(i+19,1)];
    y0_av_matrix(i,1) = sum(y0(:,i))/N;      %The matrix of an
                                             %increasing
                                             %size;
    %The average of 20 elements for each i. It is calculated
    %after the loop is finished;

    t0_av_matrix(i,1) = sum(t0(:,i))/N;      %The matrix of an
                                             %increasing
                                             %size;

    std_dev0_matrix(i,1) = std(y0(:,i),1);   %The matrix of an
                                             %increasing
                                             %size;
    %Each value is calculated from 20 points;
    std(X,1) = sqrt [(1/N)*(sum((xi-xav)^2))];

    AV_STD_DEV0_matrix(i,1) = mean(std_dev0_matrix); %This is
    %the matrix: the average for each i, from the std. dev. matrix;
    %the last value is from all the points and compared with the
    %'while' condition;

    e = size(AV_STD_DEV0_matrix);            %(The size,1);
    f = e(1,1);                              %The size;
    AV_STD_DEV0 = AV_STD_DEV0_matrix(f,1);   %Comparison with
                                             %the 'while'
                                             %condition;
end                         %THE END OF 'FOR' LOOP 1;

%THE CALCULATED PARAMETERS;
std00 = std_dev0_matrix;
t00 = t0_av_matrix;
AV00 = num2str(AV_STD_DEV0);

clear y0;
clear t0;
clear y0_av_matrix;
clear t0_av_matrix;
```

```
clear std_dev0_matrix;
clear AV_STD_DEV0_matrix;

%THE MAIN LOOP
G = 0;                      %The constant for the smoothness
                            %(critical) condition;
IT0 = 1;                    %The constant to count the number of
                            %performed iterations;
Const = 18;                 %The number of additional iterations after
                            %the critical condition is reached;
Const2 = 0;                 %The dummy constant; the real Const2
                            %is defined later; Const2 is the number of
                            %additional iterations after pressing a
                            %key to continue the calculation loop;

%The suggested values;
%400 kHz file: Const = 1, Const2 = 1 (quick);
%200 kHz file: Const = 10, Const2 = 3 (1 day);
%100 kHz file: Const = 18, Const2 = 3 (2 days);
%After 40 iterations, Smoothing_fac becomes 1, G = 1;
%After 58 iterations, Smoothing_fac becomes 1;
%G = 18 and Const = 18, G == Const (the program stops here,
%it runs only for G < Const);

for i = 1:100;              %'FOR' LOOP 2;
G;                          %Update G;
  if G == Const;            %'IF'CONDITION 1;
  %The pause is turned off;
  close (fig1);
  close (fig2);
  filename_to_del = [p,'\','AVRGED_TRIAL',filename0];
  delete (filename_to_del);
  %Const 2 is defined after the pause is done;
  G = Const-Const2;         %It will activate a few more iterations each
                            %time a key is pressed (see at the end of
                            %the 'if' loop);
  end                       %THE END OF 'IF' CONDITION 1;

%If the pause condition is not activated, the calculation process
%is going into the 'while' loop;
```

```
%The suggested value of Const = 10 for the 100 kHz file, G < 10 (10
%extra iterations will run after averaging is done);

%The program is automatically stopped, if the number of data
%points in the input file is lower than the assigned number of data
%points for averaging (d < 0); the 'for' iteration loop in the
%program runs only for d > 0 and until the critical condition for
%smoothness of the averaged curve is fulfilled (as below);

 while d > 0 & G < Const;      %'WHILE' CONDITION 1;
 AV_STD_DEV0
 d = d;                        %Update d, it is related to the
                               %file size;
 g = G                         %Update g;
 iteration_no = IT0            %The constant for counting
                               %performed iterations;

   for i = 1:d                 %'FOR' LOOP 3;
   %d is the number of points in the file, reduced by a certain
   %number of data points due to averaging; for example, the file
   %containing 500 points will have y0 (:,1), …, y0(:,481); the
   %481st column contains values from 481 to 500; the average
   %is performed for 20 consecutive values, 19 values overlap from
   %the previous dataset and 1 is new;

   y0(:,i) = [ampl0_ind(i,1) ampl0_ind(i+1,1) ampl0_ind(i+2,1)
   ampl0_ind(i+3,1) ampl0_ind(i+4,1) ampl0_ind(i+5,1)
   ampl0_ind(i+6,1) ampl0_ind(i+7,1) ampl0_ind(i+8,1)
   ampl0_ind(i+9,1) ampl0_ind(i+10,1) ampl0_ind(i+11,1)
   ampl0_ind(i+12,1) ampl0_ind(i+13,1) ampl0_ind(i+14,1)
   ampl0_ind(i+15,1) ampl0_ind(i+16,1) ampl0_ind(i+17,1)
   ampl0_ind(i+18,1) ampl0_ind(i+19,1)];

   %t0 (:,1)... t0(:,481);
   t0(:,i) = [time0_ind(i,1) time0_ind(i+1,1) time0_ind(i+2,1)
   time0_ind(i+3,1) time0_ind(i+4,1) time0_ind(i+5,1)
   time0_ind(i+6,1) time0_ind(i+7,1) time0_ind(i+8,1)
   time0_ind(i+9,1) time0_ind(i+10,1) time0_ind(i+11,1)
   time0_ind(i+12,1) time0_ind(i+13,1) time0_ind(i+14,1)
```

```
    time0_ind(i+15,1) time0_ind(i+16,1) time0_ind(i+17,1)
    time0_ind(i+18,1) time0_ind(i+19,1)];

    y0_av_matrix(i,1) = sum(y0(:,i))/N;   %The matrix of an
    %increasing size; the average of the sum of 20 elements for
    %each i; after it is finished, the loop size is (the size of
    %file,1);

    t0_av_matrix(i,1) = sum(t0(:,i))/N;   %The matrix of an
    %increasing size

    std_dev0_matrix(i,1) = std(y0(:,i),1);%The matrix of an
    %increasing size;

    %Each value is calculated from 20 points;
    std(X,1) = sqrt [(1/N)*(sum((xi-xav)^2))];

    AV_STD_DEV0_matrix(i,1) = mean(std_dev0_matrix);
    %This is the matrix average for each i, from the std. dev. matrix;
    %the last value is from all points and is compared with the
    %'while' condition;

    e = size(AV_STD_DEV0_matrix);          %(The size,1);
    f = e(1,1);                            %The size;

    AV_STD_DEV0 = AV_STD_DEV0_matrix(f,1);%Compared with
                                           %the 'while' condition;
    end             %THE END OF 'FOR' LOOP 3;

%THE CALCULATED PARAMETERS;
ampl0_ind = y0_av_matrix;                 %Get the average amplitude
                                          %for the next
                                          %averaging iteration;

time0_ind = t0_av_matrix;
AV_STD_DEV00 = AV_STD_DEV0;

%Avoid overwriting the standard deviation matrix;
std0 = std_dev0_matrix;       %This is an increasing matrix (the
                              %matrix is needed only from the last
                              %iteration step);

e1 = size(std0);
f1 = e1(1,1);
```

```
    if f < f1           %'IF' CONDITION 2;
    std0 = [];
    clear std0;
    std0(:,1) = std_dev0_matrix;
    end                 %THE END OF 'IF' CONDITION 2;

a = size(y0_av_matrix);
b = a(1,1)
d = b-N+1

    if d < 0            %'IF' CONDITION 3;
    disp('Too low number of data points for the calculation of the
    specified standard deviation')
    end                 %THE END OF 'IF' CONDITION 3;

%CHECKING THE CURVE SMOOTHNESS;
%Checking the condition for smoothness of the averaged curve,
%consisting of almost all wave points (reduced by a few points) is
%performed;
%Getting the position of the start data point (H0), by finding the
%local maximum position (within the small time interval of 1-2
%cycles determined by Help2b), then, the start point is one point
%further: +1;
%The stop point (H) is defined by finding the slope of the curve
%(corresponding to the local minimum); it is located 19 points
%before the minimum;
%The positions of H0 and H depend on how many points are per
%cycle for given file;

Help1 = size(ampl0_ind)
Help2 = Help1(1,1)
Help2b = Help2/nc    %nc is defined earlier;
Help2b = ceil(Help2b)
LocalMax = max(ampl0_ind(1:Help2b,1))%Find the local maximum, in
                     %the time interval of 1-2 cycles;
clear I;
clear J;
[I,J] = ind2sub(size(ampl0_ind),find(ampl0_ind == LocalMax)) %Find
                     %the position of the local max.
                     %I = position of max., J = 1; many
                     %positions may be found, therefore
                     %I(1,1) is taken;
H0 = I(1,1)+1        %The start point;
H00 = H0;
clear I;
clear J;
x_slope = ceil(0.1*(m/nc0))   %The number of points to check the
```

```
                              %slope of the curve =
                              %round(10%*(the file
                              %size/the number of cycles));
                              %'IF' CONDITION 4;
     if d > 0                 %Calculate the voltage difference
                              %between two neighboring points;
     diff_loop_stop = 1;      %The supportive constant;
     W0 = 1;
                              %'WHILE' CONDITION 2;
     while diff_loop_stop > 0 %Iterate until the curve changes
                              %the slope;
     k = H0;                  %The start point;
     w = W0;
     diff(w,1) = ampl0_ind(k,1)-ampl0_ind(k+1,1);
                              %The calculation of
                              %the voltage difference; there are
                              %always positive values, when the
                              %curve is smooth;
     diff_loop_stop(w,1) = ampl0_ind(k,1)-ampl0_ind(k+x_slope,1);
                              %Check the slope of the curve by
                              %comparing 2 points separated by
                              %the number of points 'x_slope';
                              %if this is negative, then the loop
                              %is stopped;
     time_diff(w,1) = time0_ind(k,1);
     H0 = H0+1;
     W0 = W0+1;
     diff_loop_stop;
     end                      %THE END OF 'WHILE' CONDITION 2;

     HelpI = size(diff)
     H = HelpI(1,1);
     H = H+H00-1;             %The stop point;
     end                      %THE END OF 'IF' CONDITION 4;

%THE CALCULATED PARAMETERS;
%diff;
%time_diff;

%Checking the sign of the voltage (amplitude) differences;
Z1= diff(:,1) > 0
Z2= diff(:,1) < 0
```

```
                        %'IF' CONDITION 5;
   if Z1(:,1) == 1 | Z2(:,1) == 0 %| means 'or';
   Smoothing_fac = 1;
   g = g+1;
   else
   Smoothing_fac = 0;
   g = 0;
   end                  %THE END OF 'IF' CONDITION 5;

%THE CALCULATED PARAMETERS;
Smoothing_fac
%g;
G = g                   %When the condition of
                        %Smoothing_fac = 1 is fulfilled, then
                        %one more extra or none (for example,
                        %one for G < 2, zero for G < 1)
                        %iterations are added; increasing
                        %the number of iterations is defined by
                        %G;

%Create the 'time' and 'diff' matrixes for plotting;
                        %'IF' CONDITION 6;
   if iteration_no == 1 %From the 1st iteration;
   diff_matrixA(:,1) = diff;
   time_diff_matrixA(:,1) = time_diff;
   end                  %THE END OF 'IF' CONDITION 6;
                        %'IF' CONDITION 7;
   if G == Const        %From the last iteration;
   diff_matrixB(:,1) = diff;
   time_diff_matrixB(:,1) = time_diff;
   end                  %THE END OF 'IF' CONDITION 7;

%THE END OF THE CURVE SMOOTHNESS TEST;
IT0 = iteration_no+1;

clear Smoothing_fac;
clear diff;
clear g;
clear y0;
clear t0;
clear y0_av_matrix;
clear t0_av_matrix;
clear std_dev0_matrix;
clear AV_STD_DEV0_matrix;
```

```
clear w;
clear W0;
clear diff;
clear time_diff;
clear diff_loop_stop;
clear iteration_no;
clear Help1
clear Help2
clear Help2b
clear LocalMax
clear H0
clear H00
clear HelpI
clear H
 end                             %THE END OF 'WHILE' CONDITION 1;

%THE CALCULATED PARAMETERS;
%G;
%IT0;
diff_matrixA;
diff_matrixB;
time_diff_matrixA;
time_diff_matrixB;
%A comment: plotting only the first and the last curve smoothness
%results;

diff1= diff_matrixA;
size_diff1 = size(diff1)
diff2 = diff_matrixB;
size_diff2 = size(diff2)
k_time1 = time_diff_matrixA;
size_k_time1 = size(k_time1)
k_time2 = time_diff_matrixB;
size_k_time2 = size(k_time2)

ampl0_ind;                       %Get the average amplitude and time;
time0_ind;
AV0 = num2str(AV_STD_DEV00); %Get the average amplitude of the
                                 %standard deviation;
std_dev0_ind = std0;

%SAVING THE AVERAGED DATA;
%datafile = [];
%filename0 = [pre_0,name]
%savefile = [p,'\','AVRGED_TRIAL',filename0] %p is defined earlier;
```

```
%datafile = [time0_ind ampl0_ind];
%save(savefile, 'datafile','-ASCII');

%PLOTTING;
fig1 = figure(1);
subplot(3,1,1);
h = plot(time0,ampl0,'-blue',time0_ind,ampl0_ind,'-cyan');
get (gca);
set(gca,'fontsize',16);
h_legend = legend('Original signal','Averaged signal');
set(h_legend,'FontSize',16);
xlabel('Time (s)','fontsize',18);
ylabel('Amplitude (V)','fontsize',18);
title(['The acquired signal',pre_0],'fontsize',20);

subplot(3,1,2)
h = plot(time0_ind,ampl0_ind,'-cyan');
get (gca);
set(gca,'fontsize',16);
title(['The acquired signal ',pre_0],'fontsize',12,'fontweigh
t','b');
h_legend = legend('Averaged signal');
set(h_legend,'FontSize',16);
xlabel('Time (s)','fontsize',12,'fontweight','b');
ylabel('Amplitude (V)','fontsize',12,'fontweight','b');

subplot(3,1,3);
h = plot(t00,std00,'-blue',time0_ind, std_dev0_ind,'-cyan');
get (gca);
set(gca,'fontsize',16);
title('Standard deviation of the signal','fontsize',12,'fontwei
ght','b');
h_legend = legend(['Original signal;','Av. std dev = ',AV00],
['Averaged signal;','Av. std dev = ',AV0]);
set(h_legend,'FontSize',16);
xlabel('Time (s)','fontsize',12,'fontweight','b');
ylabel('Amplitude (V)','fontsize',12,'fontweight','b');

fig2 = figure(2);
subplot(2,2,1);
h = plot(k_time1,diff1,'*-');
get (gca);
set(gca,'fontsize',16);
h_legend = legend(pre_0);
set(h_legend,'FontSize',16);
```

```
xlabel('Time (s)','fontsize',16);
ylabel('The amplitude difference between 2 consecutive points
(V)','fontsize',16);
title('The curve smoothness factor after the 1st
iteration','fontsize',18);
grid on;

subplot(2,2,3);
h = plot(k_time2,diff2,'*-');
get (gca);
set(gca,'fontsize',16);
h_legend=legend(pre_0);
set(h_legend,'FontSize',16);
xlabel('Time (s)','fontsize',16);
ylabel('The amplitude difference between 2 consecutive points
(V)','fontsize',16);
title('The curve smoothness factor after the last
iteration','fontsize',18);
grid on;

subplot(2,2,[2 4]);
h = plot(time0,ampl0,'-oblue',time0_ind,ampl0_ind,'->cyan');
get(gca);
set(gca,'fontsize',16);
h_legend = legend('Original signal','Averaged signal');
set(h_legend,'FontSize',16);
xlabel('Time (s)','fontsize',16);
ylabel('Amplitude (V)','fontsize',16);
title(['The acquired signal ',pre_0],'fontsize',18);
grid on;

disp('')
disp('The figure will close automatically, if any key is
pressed.')
disp('After the key pressing, a few more iterations will be run
to improve averaging')
disp('_')
disp('If averaging is OK and you want to terminate the
loop, then')
disp('saving figures in the fig format, saving the processed data
file using a different name are wise steps to do. Then press:
Ctrl+C')
disp('_')
pause;
clear Const2;
```

```
disp('Const2 is the number of extra iterations, which will be
performed')
disp('after pressing any key to continue the loop. Const2 = 3 by
default.')
disp('Const2 must be < Const.')
disp('_')
Const2 = input('Write a number for Const2:');

clear diff_matrixB;
clear time_diff_matrixB;
clear diff2;
clear k_time2;
clear AV0;
clear AV_STD_DEV00;
clear std_dev0;
clear std_dev0_ind;

%THE CALCULATED PARAMETERS:
%G;
%ampl0_ind;
%time0_ind;

end                          %THE END OF 'FOR' LOOP 2;
buffer = [];
```

5.1.5 The Cross-Correlation I

The cross-correlation of time-domain signals is the method of superposing two intervals, to determine the difference between them. The cross-correlation method described here is based on the manual choice of *t1* and *t2*, for an interval within the first signal cycle. Two intervals are correlated, the first one is from the first wave cycle, and the second one is from the second wave cycle.

```
%
buffer = [];
clear all;
format short e

path(path,'I:\...\...')
filename = 's2_low_fr_particl_PE_5kHz_vel1_1.lvm'

A = textread(filename,'',-1,'delimiter','\t','emptyvalue',NaN);
Freq = 200e6;
```

```
PRF = 5e3;                          %Hz, the wave frequency;

%time = A(1:80000,1)*(1/Freq);  %80000 points correspond to
                                    %2 cycles;
%ampl = A(1:80000,2);

time = A(:,1)*(1/Freq);             %80000 points correspond to
                                    %2 cycles;
ampl = A(:,2);
ampl = ampl+1;

c_water = 1500;                     %The sound speed in water (m/s);

figure(1)
subplot(2,1,1);
plot(time(1:40000,1),ampl(1:40000,1),time(1:40000,1),
ampl(40001:80000,1)+0.25,'linewidth',2);
get(gca)
set(gca,'fontsize',24)
title('The original signal: the 1st and the 2nd
cycles','fontsize',24)
legend('The 1st cycle','The 2nd cycle (shifted)','fontsize',16)
axis([2e-5 2.3e-5 1 1.6])
xlabel('Time (s)','fontsize',24)
ylabel('Amplitude (V)','fontsize',24)
grid on

%THE TIME AND AMPLITUDE INTERVALS;

%SIGNAL 1 (the 1st interval);
t1 = 2.04e-5;                       %Manually chosen t1 and t2 of the
                                    %interval;
t2 = t1+30*5e-9;

DSx = (t2-t1)/(1/Freq)              %The x-sample interval,
                                    %1480 points;


n1a = t1/(1/Freq)
n1b = n1a+DSx                       %The end of the 1st time
                                    %interval;

n1b-n1a
```

```
int(:,1) = ampl(n1a:n1b,1);
int_time(:,1) = time(n1a:n1b,1);

%SIGNAL 2 (the 2nd interval);
t1_y = t1+(1/PRF)                       %t1 plus the wave period
                                        %DT=1/PRF;
n1a_y = t1_y/(1/Freq)

a1 = size(time)
a2 = a1(1,1)
t_last = time(a2,1)                     %The last t;

n1b_y = t_last/(1/Freq)

int_y(:,1) = ampl(n1a_y:n1b_y,1);      %The whole part of the
                                        %signal that is left;
int_time_y(:,1) = time(n1a_y:n1b_y,1);

figure(1)
subplot(2,1,2);
plot(int_time, int,'-','linewidth',2)
get(gca)
set(gca,'fontsize',24)
legend('Subinterval for cross-correlation ','fontsize',16)
axis([2e-5 2.3e-5 1 1.3])
xlabel('Time (s)','fontsize',24)
ylabel('Amplitude (V)','fontsize',24)
grid on

figure(2)
plot(int_time, int,'-',int_time_y-(1/PRF),int_y,'-',
'linewidth',2)
legend('Interval A','Interval B (shifted)')
%axis([0 2e-4 -0.1 0.25])
xlabel('Time (s)')
ylabel('Amplitude (V)')
grid on

b1 = size(int_y)
b2 = b1(1,1)/DSx

DSx
tau0 = 0;
```

```
%tau_last = (round(b2)-1)*DSx %This equation gives a good result;
tau_last = 8e-7/(5e-9);        %For tau max. = 6e-7 s, the maximum
                               %velocity is 2.25 m/s;

%THE CROSS-CORRELATION PROCESSING;
for k = 1:tau_last
 tau = tau0;                   %THE SECOND LOOP (the
                               %outside loop);

 for i = 1:1:DSx;              %THE FIRST LOOP (the inside loop);
 CCF1(i,1) = int(i,1)*int_y(i+tau,1); %CCF1 is based on the signal
                               %multiplication;
 CCF2(i,1) = abs(int(i,1)-int_y(i+tau,1)); %CCF2 is based on
                               %the signal subtraction;
 end

CCF_sum1(k,1) = sum(CCF1);     %The number of points = Tau+1;
tau_true(k,1) = tau*(1/Freq);

CCF_sum2(k,1) = sum(CCF2);     %The number of points = Tau+1;
tau0 = tau+1;

end

CCF_sum1;
CCF_sum2;
tau_true;

figure(3)
subplot(2,1,1)
plot(tau_true(:,1),CCF_sum2(:,1),'-','linewidth',2)
get(gca)
set(gca,'fontsize',24)
legend('CCF1(i,1) = sum[abs(v1(i,1)-v2(i+tau,1))],
V','fontsize',16)
xlabel('Tau = Delta t (s)','fontsize',24)
ylabel('CCF1','fontsize',24)
grid on

subplot(2,1,2)
plot(tau_true(:,1),CCF_sum1(:,1),'-','linewidth',2)
get(gca)
set(gca,'fontsize',24)
```

```
legend('CCF2(i,1) = sum[v1(i,1)*v2(i+tau,1)]','fontsize',16)
xlabel('Tau = Delta t (s)','fontsize',24)
ylabel('CCF2','fontsize',24)
grid on
DSx
```

5.1.6 The Cross-Correlation II

The cross-correlation of time-domain signals is based on superposing two intervals, in order to determine the difference between them. The cross-correlation method described here is based on the automated selection of multiple intervals. More information can be found in Sect. 3.4.4.

The calculation procedure is the following:

- Use the acquired signal's part without wall reflexes, for example, from 2.6e-5 to 9.54e-5 s (13,880 points).
- Divide the signal into 200-point (1e-6 s) intervals, *i*.
- Start the "for" loop:
  - Find the local minimum in the interval *i*.
  - Create new intervals around the local minimum, consisting of 30 points. These intervals will be used for the cross-correlation.
  - Cross-correlate the signals using the function: *CCF* = *sum*[*v1(j)* – *v2(j* + *tau)*], where *j* is the point number and *tau* is the time shift.
  - Collect the amplitudes from the cross-correlation into a matrix, for *tau* up to 2e-6 s. Create another matrix for the frequency.
  - Plot the CCF result using the "hold on" function.
- End the "for" loop.
- Check visually if the answer is correct.

The number of the 30-point intervals, *g0*, is set to 7 by default, because there are 7 colors automatically available (enough for the analysis). The maximum number of intervals is 69.

```
%
buffer = [];
clear all;
format long e
path(path,'I:\....\....')
filename = 's2_low_fr_particl_PE_5kHz_vel1_1.lvm'
%filename = 's2_low_fr_particl_PE_5kHz_vel2_2.lvm'
%filename = 's2_low_fr_particl_PE_5kHz_vel3_2.lvm'
%filename = 's2_low_fr_particl_PE_5kHz_vel3p5_1.lvm'
%filename = 's2_low_fr_particl_PE_5kHz_vel4_1.lvm'

A = textread(filename,'',-1,'delimiter','\t','emptyvalue',NaN);
```

```
Freq = 200e6;                  %The sampling frequency, Samples/s;
PRF = 5e3;                     %The pulse repetition frequency, Hz;

time = A(:,1)*(1/Freq);        %80000 points correspond to
                               %2 cycles;

ampl = A(:,2);
ampl = ampl+1;

figure(1)
subplot(2,1,1);
plot(time(1:40000,1),ampl(1:40000,1),time(1:40000,1),
ampl(40001:80000,1)+1);
legend('The 1st cycle','The 2nd cycle, shifted')
%axis([0 2e-4 -0.5 1.5])
xlabel('Time (s)')
ylabel('Amplitude (V)')
grid on

%THE TIME AND AMPLITUDE INTERVALS;
%SIGNAL 1;

%Creating the 200-point intervals;
C0 = 0;
t1_ref = 2.6e-5                %2.6e-5 is the beginning of the first
                               %200-point interval;

%Input the number g0 of 200-point intervals for
%cross-correlation;
g0 = 2;                        %g0 = 7 is by default, g0 = 69 is the
                               %maximumdefault(1380points/200points
                               %= 69.4);

for g = 1:g0
C = C0;
T(g,1) = t1_ref+C;             %t1;
T(g,2) = T(g,1)+200*(1/Freq); %t2, for checking;
C0 = C+200*(1/Freq)+5e-9;
end

%The output parameters;
T                              %The matrix, T, consists of the
                               %beginning times t1 and the end
                               %times t2 of several intervals;
```

```
T_no_matrix = T/((1/Freq));
T_no_matrix = round(T_no_matrix)
                      %Get the indexes (the positions)
                      %instead of the times;

%Get the amplitude values of the 200-point intervals;
for g = 1:g0
  for h = 2:200

  AMPL_SUBINT(1,g) = ampl(T_no_matrix(g,1));
  %The number (the index) t1 is the beginning of each amplitude
  %interval (inup to 69 columns);

  AMPL_SUBINT(h,g) = ampl(T_no_matrix(g,1)+(h-1));
  %The first number t1+1. These are succeeding amplitude values growing
  %up to 200;

  %Get the corresponding time;
  TIME_SUBINT(1,g) = time(T_no_matrix(g,1));
  TIME_SUBINT(h,g) = time(T_no_matrix(g,1)+(h-1));
  end
end

%The output parameters;
AMPL_SUBINT;          %The amplitude values;
size(AMPL_SUBINT)     %200 points, up to 69 columns,
                      %for example, AMPL_SUBINT(200,1) should
                      %be equal to 'ampl(T_no_matrix(1,2))';
TIME_SUBINT;

clear T_no_matrix
clear T
clear A

%Get the minimum values in each of the 200-point intervals; they
%will be used for creating new 30-point intervals (for the
%cross-correlation);
for g=1:g0
min_AMPL_SUBINT(1,g) = min(AMPL_SUBINT(:,g));
end
```

```
%The output parameters;
min_AMPL_SUBINT;          %The minimum values in up to 69 columns;
size(min_AMPL_SUBINT)

%Get the position of the minimum for each interval and make a shift
%to account for the beginning of the 30-point interval, start before
%the minimum;
for g = 1:g0
  [I,J] = ind2sub(size(AMPL_SUBINT),find(AMPL_SUBINT(:,g) ==
  min_AMPL_SUBINT(1,g)));
  I = I-14;                 %The beginning of the 30-point interval
                            %(the index); there is the shift to the
                            %left by 14 points from the detected
                            %minimum;
  help = size(I);           %Avoid a double matrix;
  help2 = help(1,1);

  if help2 > 1
  I = I(1,1);
  end
  if I <= 0                 %Avoid negative index numbers;
  I = 1;
  end
  if I > 171                %Avoid the far-ends of intervals;

  I = 171;
  end

I_matrix(1,g) = I;
end

%The output parameters;
I_matrix
size(I_matrix)

clear min_AMPL_SUBINT


%Get the amplitude and time values of the 30-point (or shorter due to
%practical applications) intervals in up to 69 columns;
for g = 1:g0               %The maximum g = 1:69;
```

```
 for h = 2:24              %The 30-point new intervals; in this
                           %specific case, slightly shorter
                           %intervals are chosen instead, due to
                           %the practical application;

 NEW_AMPL_SUBINT(1,g) = AMPL_SUBINT((I_matrix(1,g)),g);
 %For g = 1, NEW_AMPL_SUBINT(1,1) = AMPL_SUBINT(I_matrix(1,1),1);
 %This is the first number for each of up to 69 columns;

 NEW_AMPL_SUBINT(h,g) = AMPL_SUBINT((I_matrix(1,g)+(h -1)),g);
 %The succeeding amplitude values, the growing matrix;

 %The corresponding time;
 NEW_TIME_SUBINT(1,g) = TIME_SUBINT((I_matrix(1,g)),g);
 NEW_TIME_SUBINT(h,g) = TIME_SUBINT((I_matrix(1,g)+(h-1)),g);
 end

figure(1)
subplot(2,1,2);
plot(NEW_TIME_SUBINT,NEW_AMPL_SUBINT,'linewidth',2);
get(gca);
set(gca,'fontsize',28);
legend('Subintervals')
%axis([0 2e-4 -0.5 1.5])
xlabel('Time (s)','fontsize',28)
ylabel('Amplitude (V)','fontsize',28)
grid on
hold on                    %Ongoing plotting for each g in the loop;

end


%The output parameters;
NEW_AMPL_SUBINT            %The amplitude of the short subintervals
                           %(30-point or shorter) in up to 69 columns;
help3 = size(NEW_AMPL_SUBINT)
help4 = help3(1,1)

NEW_TIME_SUBINT;           %The corresponding time values;
```

```
clear AMPL_SUBINT
clear I_matrix

figure(2)
subplot(2,1,1);
start_t = (t1_ref/(1/Freq));
start_a = (t1_ref/(1/Freq))+(1/PRF)/(1/Freq);
plot(time(start_t:40000,1),ampl(start_a:80000,1),'-green',
'linewidth',2);
get(gca);
set(gca,'fontsize',28);
title('Time signal','fontsize',28)
legend('The 2nd cycle incl. t1_ref, shifted')
%axis([0 2e-4 -0.5 1.5])
xlabel('Time (s)','fontsize',28)
ylabel('Amplitude (V)','fontsize',28)
grid on

%THE MAIN CROSS-CORRELATION LOOP;
%Get the 2nd part of the signal that will be used for the
%cross-correlation;
for g = 1:g0
g
t1 = NEW_TIME_SUBINT(1,g)
%t2 = NEW_TIME_SUBINT(help4,g) %t2 is not needed, just for
                              %checking;
DSx = help4                   %The interval consisting of 30 or
                              %less points;
int(:,1) = NEW_AMPL_SUBINT(:,g);
int_time(:,1) = NEW_TIME_SUBINT(:,g);

%SIGNAL 2;
t1_y = t1+(1/PRF)             %t1_y is t1 shifted by 1/PRF;
n1a_y = t1_y/(1/Freq)
n1a_y = round(n1a_y)

a1 = size(time)
a2 = a1(1,1)
t_last=time(a2,1)             %The last t;

n1b_y = t_last/(1/Freq)


ampl( n1a_y,1)
ampl( n1b_y,1)
```

```
time( n1a_y,1)
time( n1b_y,1)

int_y(:,1) = ampl(n1a_y:n1b_y,1);
                        %The part of the signal that is left;
int_time_y(:,1) = time(n1a_y:n1b_y,1);

%figure(2)
subplot(2,1,2);
plot(int_time, int,'-',int_time_y-(1/PRF),int_y+1,'-')
title('Subintervals for cross-correlation, from the 1st and 2nd
cycles')
legend('Subinterval A','Subinterval B - shifted')
%axis([0 2e-4 -0.1 0.25])
grid on

%THE CROSS_CORRELATION WITH TAU;
tau0 = 0;

%The index of tau_last;
tau_last = 500          %This is a fixed number, because
                        %tau_max = 2e-6 s is needed; for example,
                        %the expected flow velocity U is 2 m/s.
                        %DL = c*0.5*tau_max = 1500*0.5*2e-6
                        % = 1.5e-3 m, U = DL*PRF = 7.5 m/s;

for k = 1:tau_last      %THE SECOND LOOP;
tau = tau0;

 for I = 1:1:DSx;       %THE FIRST LOOP;
 CCF(i,1) = abs(int(i,1)-int_y(i+tau,1));
 end                    %THE END OF THE FIRST LOOP;

%The output parameters;
CCF_sum(k,1) = sum(CCF); %The number of points is tau+1;
tau_true(k,1) = tau*(1/Freq);
tau0 = tau+1;

end                     %THE END OF THE SECOND LOOP;


%The output parameters;
CCF_sum(:,g) = CCF_sum;
tau_true(:,g) = tau_true;
```

```
CCF_final(:,1) = tau_true(1:400,1); %tau is only from 0 to 2e-6 s
CCF_final(:,g+1) = CCF_sum(1:400,1);

figure(3)
subplot(2,1,1)
plot(tau_true,CCF_sum,'linewidth',2)
%plot(tau_true(1:400,g),CCF_sum(1:400,g)) %The optional plotting;
axis([0 1e-6 0 3])
get(gca);
set(gca,'fontsize',28);
legend('CCF = sum[abs(v1(j)-v2(j+tau))]')
title('Cross-Correlation','fontsize',28)
xlabel('Tau = Delta t (s)','fontsize',28)
ylabel('CCF (V)','fontsize',28)
grid on
hold on

clear CCF_sum
clear tau_true
clear int
clear int_time
clear int_y
clear int_time_y
clear t1
clear n1_a_y
clear n1_b_y

end                    %THE END OF THE MAIN LOOP;


%The output parameters;
CCF_final
size(CCF_final)        %The size is 400x70. The first column is for
                       %tau (s), the other columns are the CCF (V);
```

5.1.7 The Signal Amplitude Stretching Algorithm

The development of an algorithm for signal amplitude stretching was driven by the need of improving the quality of signals acquired for cross-correlation processing. The signals are used for the determination of flow velocity. The algorithm is presented here and more information can be found in Sect. 3.4.5.

```
%
buffer = [];
clear all;
format long e

p = 'I:\...\...';                %The source path;
p2 = 'I:\...\...';               %The file saving path;

path(path, p);
%filename = 's2_low_fr_particl_PE_5kHz_vel1_1.lvm'
%filename = 's2_low_fr_particl_PE_5kHz_vel2_2.lvm'
filename = 's2_low_fr_particl_PE_5kHz_vel3p5_1.lvm'
%filename = 's2_low_fr_particl_PE_5kHz_vel4_1.lvm'

A = textread(filename,'',-1,'delimiter','\t','emptyvalue',NaN);

Freq = 200e6;                    %The sampling frequency, Samples/s;
PRF = 5e3;                       %The pulse repetition frequency, Hz;
t1 = 3000;                       %The number of points at the beginning
                                 %to skip;

%The interval from the 1st wave cycle;
time1 = A(t1:40000,1)*(1/Freq);
                                 %80000 points correspond to 2 cycles;

%The interval from the 2nd wave cycle;
time2 = A(40000+t1:80000,1)*(1/Freq);

ampl1 = A(t1:40000,2);
ampl2 = A(40000+t1:80000,2);
ampl1 = ampl1+1;
ampl2 = ampl2+1;

figure(1)
plot(time1,ampl1,time2,ampl2);
get(gca)
set(gca,'fontsize',24);
legend('The 1st cycle','The 2nd cycle','fontsize',16)
xlabel('Time (s)','fontsize',24)
ylabel('Amplitude (V)','fontsize',24)
grid on

DT = 5*1e-6*Freq                 %The number of points in the interval for
                                 %analysis (1e-6 is for 200-point
                                 %intervals);
```

```
%Choose whether the 1st or 2nd wave cycle is processed, and save
%the final processed file accordingly; both cycles must be run
%separately;

%Choose the 1st or 2nd wave cycle to be considered;
%The 1st wave cycle;
DT1 = time1(1:DT);          %The 1st interval;
DT2 = time1(DT+1:2*DT);     %The 2nd interval;
DT3 = time1(2*DT+1:3*DT);   %The 3rd interval;
AMP1 = ampl1(1:DT);
AMP2 = ampl1(DT+1:2*DT);
AMP3 = ampl1(2*DT+1:3*DT);
CYCLE_NO_FOR_SAVING = '1CYC_INT1';

%OR;

%The 2nd wave cycle;
%DT1 = time2(1:DT);
%DT2 = time2(DT+1:2*DT);
%DT3 = time2(2*DT+1:3*DT);
%AMP1 = ampl2(1:DT);
%AMP2 = ampl2(DT+1:2*DT);
%AMP3 = ampl2(2*DT+1:3*DT);
%CYCLE_NO_FOR_SAVING = '2CYC_INT1';

figure(2)
subplot(2,1,1)
plot(DT1,AMP1,DT2,AMP2,DT3,AMP3,'linewidth',2)
get(gca)
set(gca,'fontsize',24);
title('Original','fontsize',24)
legend('Part 1','Part 2','Part 3','fontsize',16)
xlabel('Time (s)','fontsize',24)
ylabel('Amplitude (V)','fontsize',24)
grid on

%The average amplitude of the selected part of the signal;
AV1 = mean(AMP1)
AV2 = mean(AMP2)
AV3 = mean(AMP3)

%"Zeroing out" (shifting) the signal;
AMP1 = AMP1-AV1;
AMP2 = AMP2-AV2;
AMP3 = AMP3-AV3;
```

```
%The maximum of the selected part of the signal (zeroed out);
MAX1 = max(AMP1)
MAX2 = max(AMP2)
MAX3 = max(AMP3)

%The minimum of the selected part of the signal (zeroed out);
MIN1 = min(AMP1);
MIN2 = min(AMP2);
MIN3 = min(AMP3);

figure(2)
subplot(2,1,2)
plot(DT1,AMP1,DT2,AMP2,DT3,AMP3,'linewidth',2)
get(gca)
set(gca,'fontsize',24);
legend('Part 1','Part 2','Part 3','fontsize',16)
title('Amplitude zeroed out','fontsize',24)
xlabel('Time (s)','fontsize',24)
ylabel('Amplitude (V)','fontsize',24)
grid on

%SIGNAL PROCESSING;
%1. THE TOP PART OF THE SIGNAL (A >= 0);
%For the selected (1st or 2nd) wave cycle interval;

%Defining the range of 3 segments for the top part of the signal;
Segm_top = (MAX1)/3

%The 1st amplitude segment, filtering out the data points;
Atop1_m = (AMP1 >= 0 & AMP1 < Segm_top %The 1-0 matrix type;

%The matrix with the values only in the determined above range;
Atop1 = AMP1(Atop1_m)

%Checking;
length(Atop1_m)
length(Atop1)

AV1+Segm_top
AV1

%The 2nd amplitude segment;
Atop2_m = (AMP1 >= Segm_top & AMP1 < 2*Segm_top);
Atop2 = AMP1(Atop2_m);

%The 3rd amplitude segment;
```

```
Atop3_m = (AMP1 >= 2*Segm_top & AMP1 <= 3*Segm_top);
Atop3 = AMP1(Atop3_m);

L1t = length(Atop1)
L2t = length(Atop2)
L3t = length(Atop3)
L = length(AMP1)

figure(3)
subplot(2,1,1)
plot((1:1:length(Atop1)),Atop1,(1:1:length(Atop2)),
Atop2,(1:1:length(Atop3)),
Atop3,'linewidth',2)
get(gca)
set(gca,'fontsize',24);
title('Selected amplitude segments for fixed amplitude
range','fontsize',24)
legend('1 segment top','2 segment top','3 segment
top','fontsize',16)
xlabel('Time (s)','fontsize',24)
ylabel('Amplitude (V)','fontsize',24)
grid on

%Generation of the stretched amplitude; the multiplication of
%three signals' segments by the individual constants C1, C2 = 2C1,
%C3 = 3C1, and then creation of the new amplitude out of the
%results;

%SEGMENT 1;
C1 = 2                              %The multiplication constant for
                                    %the 1st segment;
Atop1_m
Atop1
for i = 1:length (Atop1);
 Atop1_mM(i,1) = C1*Atop1(i,1);    %M - magnified (the amplitude with
                                    %the multiplied values);
end
Atop1_mM                            %The new matrix with
                                    %stretched values;

%SEGMENT 2;
C2 = 4                              %The multiplication constant for
                                    %the 2nd segment;
```

```
for i = 1:length (Atop2);       %i is the same as for SEGMENT 1;
  Atop2_mM(i,1) = C2*Atop2(i,1);
end
Atop2_mM                        %The new matrix with stretched values;

%SEGMENT 3;
C3 = 6                          %The multiplication constant for the 3rd
                                %segment;
for i = 1:length (Atop3);%i is the same as for SEGMENT 1;
 Atop3_mM(i,1) = C3*Atop3(i,1);
end
Atop3_mM                        %The new matrix with stretched values;

%2. THE BOTTOM PART OF THE SIGNAL (A < 0);
%For the selected (1st or 2nd) wave cycle interval;

%Define the range of 3 segments for the bottom part of the signal;
Segm_bot = (MIN1)/3;     %This is a negative number;

%The 1st amplitude segment, filter out the data points;
Abot1_m = (AMP1 < 0 & AMP1 > Segm_bot);
                                %The 1-0 matrix type;
Abot1 = AMP1(Abot1_m);

%The 2nd amplitude segment;
Abot2_m = (AMP1 <= Segm_bot & AMP1 > 2*Segm_bot);
Abot2 = AMP1(Abot2_m);

%The 3rd amplitude segment;
Abot3_m = (AMP1 <= 2*Segm_bot & AMP1 >= 3*Segm_bot);
Abot3 = AMP1(Abot3_m);

L1b = length(Abot1)
L2b = length(Abot2)
L3b = length(Abot3)

figure(3)
subplot(2,1,2)
plot((1:1:length(Abot1)),Abot1,(1:1:length(Abot2)),
Abot2,(1:1:length(Abot3)),Abot3,'linewidth',2)
get(gca)
set(gca,'fontsize',24);
legend('1 segment bottom','2 segment bottom','3 segment bottom',
'fontsize',16)
```

```
title('Selected amplitude segments for fixed amplitude
range','fontsize',24)
xlabel('Time (s)','fontsize',24)
ylabel('Amplitude (V)','fontsize',24)
grid on

%Checking;
Sum_points = L1t+L2t+L3t+L1b+L2b+L3b;
if Sum_points == L
 display ('OK')
else
 display ('not OK')
 pause on
end

%Generation of the stretched amplitude; multiplication of three
%signals' segments by the individual constants C1, C2 = 2C1,
%C3 = 3C1, and then creation of the new amplitude out of the
%results;

%SEGMENT 1;
for i = 1:length (Abot1);
  Abot1_mM(i,1) = C1*Abot1(i,1);%M - magnified (the amplitude with
                        %the multiplied values);
end
Abot1_mM                %The new matrix with stretched values;

%SEGMENT 2;
for i = 1:length (Abot2);
                        %i is the same as for SEGMENT 1;
  Abot2_mM(i,1) = C2*Abot2(i,1);
end
Abot2_mM                %The new matrix with stretched values;

%SEGMENT 3;
for i = 1:length (Abot3);
                        %i is the same as for SEGMENT 1;
  Abot3_mM(i,1) = C3*Abot3(i,1);
end
Abot3_mM                %The new matrix with stretched values;

%COMBINING ALL THE PROCESSED POINTS (THE STRETCHED
%AMPLITUDES) INTO A NEW MATRIX;
H = size(AMP1);
H1 = H(1,1)
```

```
i = [];
k0 = 1;
l0 = 1;
m0 = 1;
n0 = 1;
o0 = 1;
p0 = 1;

for i = 1:H1
 if Atop1_m(i,1) == 1
 k = k0;
 N_amp_C (i,1) = Atop1_mM(k,1)
 k0 = k + 1;
 end
 if Atop2_m(i,1) == 1
 l = l0;
 N_amp_C (i,1) = Atop2_mM(l,1)
 l0 = l + 1;
 end
 if Atop3_m(i,1) == 1
 m = m0;
 N_amp_C (i,1) = Atop3_mM(m,1)
 m0 = m + 1;
 end
 if Abot1_m(i,1) == 1
 n = n0;
 N_amp_C (i,1) = Abot1_mM(n,1)
 n0 = n + 1;
 end
 if Abot2_m(i,1) == 1
 o = o0;
 N_amp_C (i,1) = Abot2_mM(o,1)
 o0 = o + 1;
 end
 if Abot3_m(i,1) == 1
 p = p0;
 N_amp_C (i,1) = Abot3_mM(p,1)
 p0 = p + 1;
 end
 k0;
 l0;
 m0;
 n0;
 o0;
 p0;
```

```
end
N_amp_C;                    %The new matrix with the combined data
                            %points;

figure (4)
subplot(2,1,1)
plot(DT1,N_amp_C,'linewidth',2)
get (gca);
set(gca,'fontsize',24);
title('Combined magnified matrix,Part 1','fontsize',24)
xlabel('Time (s)','fontsize',24)
ylabel('Amplitude (V)','fontsize',24)
axis([DT1(1,1) DT1(H1,1) 5*MIN1 5*MAX1])
grid on

subplot(2,1,2)
plot(DT1,AMP1,'linewidth',2)
get (gca);
set(gca,'fontsize',24);
title('Original (zeroed out) signal for comparison,Part
1','fontsize',24)
xlabel('Time (s)','fontsize',24)
ylabel('Amplitude (V)','fontsize',24)
axis([DT1(1,1) DT1(H1,1) 5*MIN1 5*MAX1])
grid on

%Saving the data into a text file;
datafile = [];
savefile = [p2,'\',CYCLE_NO_FOR_SAVING,filename]
datafile = [DT1, N_amp_C];
save(savefile,'datafile','-ASCII');

display('The processing was performed for DT: ')
DT
```

5.1.8 The Fast Fourier Transform

Plotting the frequency spectra by using the fast Fourier transform (FFT) function for two time-domain wave signals is presented here. The FFT produces the frequency spectrum out of a time-domain signal.

```
%
buffer = [];
format long e

path(path,'F:\...\...')
filename = 'water_speed3_PE_8.lvm'
A = textread(filename,'',-1,'delimiter','\t', 'headerlines',0,'emp
tyvalue',NaN);
freq = 200*10^6;
time = A(:,1)*(1/freq);
size_time = size(time)
ampl = A(:,2);
N = length(ampl)

Y = fft(ampl,N);            %Converting to the frequency domain;
Pyy = Y.*conj(Y)/N;         %Converting to the power domain (power
                            %at various frequencies);
Numpsec = freq              %The sampling frequency (numbers per
                            %second);
Adr = fix(N/2)
f = Numpsec*(0:Adr)/N;
Sumcount = sum(Pyy(1:Adr+1))

filename2 = 'water_speed3_PE_beads_c4_.lvm'
A2 =
textread(filename2,'',-1,'delimiter','\t','headerlines',0,'emptyva
lue',NaN);
freq2 = freq;
time2 = A2(:,1)*(1/freq2);
size_time2 = size(time2)
ampl2 = A2(:,2);
N2 = length(ampl2)

Y2 = fft(ampl2,N2);
Pyy2 = Y2.*conj(Y2)/N2;
Numpsec2 = freq2
Adr2 = fix(N2/2)
f2 = Numpsec2*(0:Adr2)/N2;
Sumcount2 = sum(Pyy2(1:Adr2+1))

figure(1)
subplot(2,1,1);plot(time,ampl,'-oblack',time2,ampl2,'-*yellow')
title('Pulse-echo test','fontsize',14)
legend('Water flow','Water-particle flow','fontsize',14)
xlabel('Time (s)','fontsize',14)
```

```
ylabel('Amplitude (V)','fontsize',14)
grid on

subplot(2,1,2);loglog(f,Pyy(1:Adr + 1),'-oblack',f2,Pyy2(1:Adr2+1
),'-yellow')
title('Frequency spectrum','fontsize',14)
xlabel('Frequency (Hz)','fontsize',14)
ylabel('Amplitude (V)','fontsize',14)
legend('Water flow','Water-particle flow','fontsize',14)
grid on
```

5.2 A Programming Code for Image Processing

5.2.1 *The Particle Image Velocimetry*

The particle image velocimetry (PIV) program that was used for the PIV results in Sect. 3.3.6 is presented here. The MatPIV 1.6.1 package was used for the recalculation of velocities from images.

Valuable practical information regarding the code is the following: the correct velocity matrix is (u, vny); after clicking on a plotted vector arrow, the vector arrow, v, pointed downwards is negative ($vny = -v$). This applies to the software version used.

```
%
p = genpath('C:\...\...');
addpath(p);
clc
format compact

im1 = 'Test2_55rpm_0p69ms_106um particl_067.bmp';
im2 = 'Test2_55rpm_0p69ms_106um particl_074.bmp';

%Plotting the image;
picture = imread(im1);
dim = size(picture)

figure(1)
image(picture);
hold on

%Reading the images and calculating the associated speeds in
%pixels/second; The time difference between the frames = 7.8 ms;
```

```
%The effect of the overlapping of pictures;

[x,y,u,v] = matpiv(im1,im2,[64 64; 32 32; 16 16; 16
16],7.8e-3,0.5,'multin');

%Converting the picture scale to the real scale by the factor
%c (cm to px), for example, 37.01 px in a picture is 0.79 cm in
%the real world, c = 0.79/37.01; by default c = 1;

%For the liquid-particle pictures: 72 px = 1 cm;
c = 1/72;
u = u*c;
v = v*c;

%Reading the picture pixels in the x- and y-directions using
%Microsoft Picture Manager;
lx = 512;
ly = 576;
yny = dim(1)-y
vny = -v
quiv = 3
quiver(x,y,u,v,quiv,'Color','r');
hold off

figure(2);
quiver(x,yny,u,vny,quiv);      %quiver shows the velocities as
                               %the arrow vectors;
%PART 1;
%The velocity along the y-axis for the selected value of x;
%Click any marker on Fig. 1 to check j;
%Checking x0 and y0 values in the workspace;

j = 31                         %The 31st column in the matrixes;
y0 = y(:,j)                    %y0 is from 8 to 568 (px);
x0 = x(:,j)                    %x0 is 248 (px);
u0 = u(:,j)                    %u0 is the velocity vector in the
                               %x-direction;
v0 = vny(:,j)                  %v0 is the velocity vector in the
                               %y-direction;

figure(3)
```

```
subplot(2,2,[1 3]);quiver(x0,y0,u0,v0,quiv,'Color','r');
                              %The velocity values
                              %are not scaled;
legend('Velocity vectors at x = 248')
xlabel('x (px)')
ylabel('y (px)'
grid on

figure(3)
subplot(2,2,2);plot(y0,u0,'-o')
legend('Horizontal velocity along y at x = 248')
xlabel('y (px)')
ylabel('Velocity u (cm/s)')
grid on

figure(3)
subplot(2,2,4);plot(y0,v0,'-o')
legend('Vertical velocity along y at x = 248')
xlabel('y (px)')
ylabel('Velocity v (cm/s)')
grid on

%PART 2;
%The velocity along the x-axis for the selected value of y;
%Click any marker on Fig. 1 to check j;
%Checking x0 and y0 values in the workspace;

j = 4                         %The 4th column in the matrixes;
y0 = y(j,:)                   %y0 is 32 (px);
x0 = x(j,:)                   %x0 is from 0 to 504 (px);
u0 = u(j,:)                   %u0 is the velocity vector in the
                              %x-direction;
v0 = vny(j,:)                 %v0 is the velocity vector in the
                              %y-direction;

figure(4)
subplot(2,2,[1 3]);quiver(x0,y0,u0,v0,quiv,'Color','r');
                              %The velocity values
                              %are not scaled;
legend('Velocity vectors at y = 32')
xlabel('x (px)')
ylabel('y (px)')
grid on
```

```
figure(4)
subplot(2,2,2);plot(x0,u0,'-o')
legend('Horizontal velocity along x at y = 32')
xlabel('x (px)')
ylabel('Velocity u (cm/s)')
grid on

figure(4)
subplot(2,2,4);plot(x0,v0,'-o')
legend('Vertical velocity along x at y = 32')
xlabel('x (px)')
ylabel('Velocity v (cm/s)')
grid on
```

Competing Interests The author has no conflicts of interest to declare that are relevant to the content of this chapter.

Index

B. Omylska, *Ultrasonic Measurement and Characterization of Liquid-Particle Flow*, https://doi.org/10.1007/978-3-032-03957-6

The manufacturer's authorised representative in the EU is Springer Nature Customer Service Centre GmbH, Europaplatz 3, 69115 Heidelberg, Germany. If you have any concerns regarding our products, please contact ProductSafety@springernature.com

Printed and bound by CPI Group (UK) Ltd, Croydon, CR0 4YY
17/09/2026
02218458-0001